Recent Trends in Phosphate Mining and Beneficiation and Related Waste Management

Recent Trends in Phosphate Mining and Beneficiation and Related Waste Management

Special Issue Editors

Mostafa Benzaazoua
Yassine Taha

MDPI • Basel • Beijing • Wuhan • Barcelona • Belgrade

Special Issue Editors
Mostafa Benzaazoua
University of Quebec
Canada

Yassine Taha
Mohammed VI Polytechnic University
Morocco

Editorial Office
MDPI
St. Alban-Anlage 66
4052 Basel, Switzerland

This is a reprint of articles from the Special Issue published online in the open access journal *Minerals* (ISSN 2075-163X) in 2019 (available at: https://www.mdpi.com/journal/minerals/special_issues/phosphate_days).

For citation purposes, cite each article independently as indicated on the article page online and as indicated below:

LastName, A.A.; LastName, B.B.; LastName, C.C. Article Title. *Journal Name* **Year**, *Article Number*, Page Range.

ISBN 978-3-03928-172-5 (Pbk)
ISBN 978-3-03928-173-2 (PDF)

© 2020 by the authors. Articles in this book are Open Access and distributed under the Creative Commons Attribution (CC BY) license, which allows users to download, copy and build upon published articles, as long as the author and publisher are properly credited, which ensures maximum dissemination and a wider impact of our publications.

The book as a whole is distributed by MDPI under the terms and conditions of the Creative Commons license CC BY-NC-ND.

Contents

About the Special Issue Editors . . . . **vii**

Preface to "Recent Trends in Phosphate Mining and Beneficiation and Related Waste Management" . . . . **ix**

Yassine Taha and Mostafa Benzaazoua
Editorial for Special Issue: "Recent Trends in Phosphate Mining, Beneficiation and Related Waste Management"
Reprinted from: *Minerals* **2019**, *9*, 755, doi:10.3390/min9120755 . . . . **1**

Yaoyang Ruan, Dongsheng He and Ruan Chi
Review on Beneficiation Techniques and Reagents Used for Phosphate Ores
Reprinted from: *Minerals* **2019**, , 253, doi: **3**

Haïfa Boujlel, Ghassen Daldoul, Haïfa Tlil, Radhia Souissi, Noureddine Chebbi, Nabil Fattah and Fouad Souissi
The Beneficiation Processes of Low-Grade Sedimentary Phosphates of Tozeur-Nefta Deposit (Gafsa-Metlaoui Basin: South of Tunisia)
Reprinted from: *Minerals* **2019**, *9*, 2, doi:10.3390/min9010002 . . . . **21**

Elves Matiolo, Hudson Jean Bianquini Couto, Michelle Fernanda de Lira Teixeira, Renata Nigri de Almeida and Amanda Soares de Freitas
A Comparative Study of Different Columns Sizes for Ultrafine Apatite Flotation
Reprinted from: *Minerals* **2019**, *9*, 391, doi:10.3390/min9070391 . . . . **36**

Xiao Sheng Yang, Hannu Tapani Makkonen and Lassi Pakkanen
Rare Earth Occurrences in Streams of Processing a Phosphate Ore
Reprinted from: *Minerals* **2019**, *9*, 262, doi:10.3390/min9050262 . . . . **47**

Shuai Li, Jie Zhang, Huaifa Wang and Caili Wang
Geochemical Characteristics of Dolomitic Phosphorite Containing Rare Earth Elements and Its Weathered Ore
Reprinted from: *Minerals* **2019**, *9*, 416, doi:10.3390/min9070416 . . . . **59**

Joseph P. Laurino, Jack Mustacato and Zachary J. Huba
Rare Earth Element Recovery from Acidic Extracts of Florida Phosphate Mining Materials Using Chelating Polymer 1-Octadecene, Polymer with 2,5-Furandione, Sodium Salt
Reprinted from: *Minerals* **2019**, *9*, 477, doi:10.3390/min9080477 . . . . **71**

Mohamed Loutou, Wafa Misrar, Mohammed Koudad, Mohammed Mansori, Liga Grase, Claude Favotto, Yassine Taha and Rachid Hakkou
Phosphate Mine Tailing Recycling in Membrane Filter Manufacturing: Microstructure and Filtration Suitability
Reprinted from: *Minerals* **2019**, *9*, 318, doi:10.3390/min9050318 . . . . **81**

Mustapha Amrani, Yassine Taha, Azzouz Kchikach, Mostafa Benzaazoua and Rachid Hakkou
Valorization of Phosphate Mine Waste Rocks as Materials for Road Construction
Reprinted from: *Minerals* **2019**, *9*, 237, doi:10.3390/min9040237 . . . . **99**

Yikai Liu, Qinli Zhang, Qiusong Chen, Chongchong Qi, Zhu Su and Zhaodong Huang
Utilisation of Water-Washing Pre-Treated Phosphogypsum for Cemented Paste Backfill
Reprinted from: *Minerals* **2019**, *9*, 175, doi:10.3390/min9030175 . . . . **114**

Xibing Li, Shitong Zhou, Yanan Zhou, Chendi Min, Zhiwei Cao, Jing Du, Lin Luo and Ying Shi
Durability Evaluation of Phosphogypsum-Based Cemented Backfill Through Drying-Wetting Cycles
Reprinted from: *Minerals* **2019**, *9*, 321, doi:10.3390/min9050321 . . . . **133**

Zhihui Shen, Qin Zhang, Wei Cheng and Qianlin Chen
Radioactivity of Five Typical General Industrial Solid Wastes and its Influence in Solid Waste Recycling
Reprinted from: *Minerals* **2019**, *9*, 168, doi:10.3390/min9030168 . . . . **146**

About the Special Issue Editors

Mostafa Benzaazoua joined the University of Quebec (UQAT) in 1996, as postdoctoral fellow. He became professor in June 1997 in the same university and held a Canada Research Chair (CRC 2003–2011) in "mine waste integrated management". In 2009, he co-chairs an International Research Chair funded by the International Development Research Centre (IDRC 2009-2014) jointly with the CRC program jointly with UCA Marrakech University (Morocco). He is presently Affiliated Professor at the University Mohamed IV Polytechnique in Morocco (UM6P) leading the program of Mining Environment and Circular Economy. He is also Visiting Professor at the Wuhan Institute of technology in China (WIT). He has worked on large number of government-funded and industry-sponsored projects, dealing with applied mineralogy and geochemistry for mine pollution control, waste management and valorization, mine site rehabilitation, and mineral processing. At the end of 2010, he took a secondment and joined the National Institute of Applied Sciences at Lyon in France as a University Professor, which allow him to diversify his research themes working on other industrial mineral waste management, environmental evaluation, treatment and reuse (dredged sediments and sewage sludges, incineration by-products, etc.). In 2012, he returned to UQAT, joining its recent Research Institute of Mining and Environment. M. Benzaazoua is familiar with various spectroscopic and microbeam-based mineralogical techniques, most mineral processing techniques, as well as lab and field physical modeling applied to study the pollution generation potential from mining and industrial wastes before and after treatment/stabilization/valorization. Presently, he is working on the adaptation of Circular Economy, Industrial Ecology and Geometallurgy to mine industry.

Yassine Taha is an Assistant Professor at Mohammed VI Polytechnic University (UM6P), Morocco. His research interests revolve around phosphate sustainable mining, mine waste management and valorization, development of high-value added materials based on industrial solid wastes, resource recovery and low-grade ore beneficiation, low carbon footprint materials for construction, life cycle assessment and circular economy.

Preface to "Recent Trends in Phosphate Mining and Beneficiation and Related Waste Management"

Phosphorus is one of the strategic and critical raw materials and is an irreplaceable essential nutrient for life for various planetary living organisms. However, global high-grade phosphate reserves are known to be decreasing, and the need to mine low-grade ores, including former waste rocks and tailings, in order to extract phosphorus-bearing minerals (i.e., apatite) is becoming increasingly crucial.

At the same time, concerns about access to critical raw materials and their availability for the future of the industry, in particular for energy storage, are growing. Nevertheless, phosphate ores are known to contain some of these elements, such as rare earth elements and uranium, that may represent valuable bonuses in phosphate ore trading. The recovery of these vital elements from phosphate waste may help in developing the needs of green energy for the future and contribute to the achievement of the sustainable development goals around the phosphate industry.

Also, various types of waste stream are continuously produced by the phosphate industry, such as carbonated and/or siliceous waste rocks during ore extraction, clayey sludges (or tailings) produced during apatite concentration, and phosphogypsum generated by the chemical industry that transforms the apatite into phosphoric acid. These waste products consititute huge volumes, reaching a ratio between 5 to 10 tons of waste per ton of concentrated phosphate apatite. The management of these waste products is becoming an important issue in terms of public concern and environmental and financial aspects. In the framework of sustainable mining and circular economy objectives, several ecofriendly and green solutions for the recycling and management of waste rocks from phosphate mining, tailings, and phosphogypsum are highlighted in this Special Issue.

The published papers in this Special Issue highlight the opportunities related to the beneficiation, management, and recycling of low-grade phosphate ores as well as the numerous related by-products considered as waste but that can be valorized/reused as secondary raw materials. .

Mostafa Benzaazoua, Yassine Taha
Special Issue Editors

Editorial

Editorial for Special Issue: "Recent Trends in Phosphate Mining, Beneficiation and Related Waste Management"

Yassine Taha [1,*] and Mostafa Benzaazoua [1,2,*]

[1] Mohammed VI Polytechnic University, Lot 660, Hay Moulay Rachid, Ben Guerir 43150, Morocco
[2] Institut de Recherche en Mines et en Environnement, Univ. du Québec en Abitibi Témiscamingue, 445 Boul de l'Université, Rouyn-Noranda, QC J9X 5E4, Canada
* Correspondence: Yassine.TAHA@um6p.ma (Y.T.); Mostafa.Benzaazoua@uqat.ca (M.B.)

Received: 26 November 2019; Accepted: 27 November 2019; Published: 4 December 2019

This Special Issue contains a series of selected papers concerning phosphate along with its mining and transformation life cycle. Phosphate extraction and beneficiation is one of the most vital mining industries in the world. Phosphorus derived from phosphate ores is a vital element of life and an exceptional component in fertilizers and food industries. However, many challenges are currently being faced in the extraction of phosphorus, one of the most strategic and critical raw materials [1]. Global high-grade phosphate reserves are known to be decreasing, and the need to explore low-grade ores, including former waste rocks and tailings, to extract apatite is becoming increasingly crucial.

In this regard, a recent review of the beneficiation techniques and reagents that can be used for the beneficiation of low-grade ores with high impurity contents was conducted by Ruan et al. [2]. The beneficiation process of a low-grade sedimentary phosphate ore (12 wt.% P_2O_5) from the Gafsa-Metlaoui sedimentary basin in southern Tunisia was investigated [3]. The researchers succeeded in reaching a recovery rate of 92.4% and improved the P_2O_5 grade of concentrate to 27.1%. Moreover, Matiolo et al. [4] investigated the possibility of recovering ultrafine apatite particles, usually lost within tailings, by comparing different column size flotation processes.

In addition, phosphate ores are known to contain other critical raw materials (CRMs), such as rare earth elements and uranium, that may represent valuable bonuses in phosphate ore trading. The recovery of these vital elements from phosphate wastes may help in developing the needs of green energy in the future and contribute to the achievement of the Sustainable Development Goals. The occurrence of rare earth elements (REEs) and their bearing phases within different streams of phosphate ore processing in China was discussed by Yang et al. [5], using mineral liberation analysis (MLA) coupled with electron probe microanalysis (EPMA). Apatite, allanite, monazite, pyrochlore, and gypsum were identified as the main REE-bearing phases in the different studied samples. Li et al. [6] investigated the geochemical and mineralogical characteristics of primary and weathered dolomitic phosphorites containing REEs from the Zhijin mining district in Guizhou Province, China. Novel polymer and chemical products were also used to recover REEs from acidic extracts of Florida phosphate mining materials [7].

Various types of waste streams are continuously produced by the phosphate industry such as carbonated and/or siliceous waste rocks, clayey sludge, and phosphogypsum. These wastes represent huge volumes, reaching a ratio between 5 and 10 tons of waste per ton of concentrated phosphate apatite. The management of these wastes is becoming an important issue in terms of public concerns and environmental and financial aspects. In the framework of sustainable mining and circular economy objectives, several ecofriendly and green solutions for the recycling and management of phosphate mine waste rocks, tailings, and phosphogypsum are highlighted in this Special Issue. Phosphate mine tailings coming from beneficiation plants were tested for their potential reuse as membrane filter products [8]. Also, it was proven by Amrani et al. [9] that phosphate mine waste rocks can be successfully used as potential alternative secondary raw materials in road construction. According to this study, the phosphate mine waste rocks were proven as natural aggregates similar to conventional materials commonly used for road construction applications. Phosphogypsum is the subject of two other papers in this Special Issue. It was demonstrated that phosphogypsum can be used in a sustainable way in cemented paste backfill applications [10,11]. The radioactivity of different industrial solid wastes, phosphogypsum among them, was assessed by Shen et al. [12]. The study provided a quantitative analysis for the safe use of the evaluated wastes in Guizhou building materials.

The published papers in this Special Issue highlight the opportunities related to the beneficiation, management, and recycling of low-grade phosphate ores as well as the numerous related by-products considered as wastes but that can be valorized/reused as secondary raw materials. These solutions can contribute to resource recovery from the growing amounts of mine wastes, finite natural resource conservation by recycling of these wastes, and environmental impact reduction.

Conflicts of Interest: The author declares no conflict of interest.

References

1. European Commission. *Communication from the Commission to the European Parliament, the Council, the European Economic and Social Committee and the Committee of the Regions on the 2017 List of Critical Raw Materials for the EU*; European Commission: Brussels, Belgium, 2017.
2. Ruan, Y.; He, D.; Chi, R. Review on Beneficiation Techniques and Reagents Used for Phosphate Ores. *Minerals* **2019**, *9*, 253. [CrossRef]
3. Boujlel, H.; Daldoul, G.; Tlil, H.; Souissi, R.; Chebbi, N.; Fattah, N.; Souissi, F. The Beneficiation Processes of Low-Grade Sedimentary Phosphates of Tozeur-Nefta Deposit (Gafsa-Metlaoui Basin: South of Tunisia). *Minerals* **2019**, *9*, 2. [CrossRef]
4. Matiolo, E.; Couto, H.J.B.; de Lira Teixeira, M.F.; de Almeida, R.N.; de Freitas, A.S. A Comparative Study of Different Columns Sizes for Ultrafine Apatite Flotation. *Minerals* **2019**, *9*, 391. [CrossRef]
5. Yang, X.; Makkonen, H.T.; Pakkanen, L. Rare Earth Occurrences in Streams of Processing a Phosphate Ore. *Minerals* **2019**, *9*, 262. [CrossRef]
6. Li, S.; Zhang, J.; Wang, H.; Wang, C. Geochemical characteristics of dolomitic phosphorite containing rare earth elements and its weathered ore. *Minerals* **2019**, *9*, 416. [CrossRef]
7. Laurino, J.P.; Mustacato, J.; Huba, Z.J. Rare Earth Element Recovery from Acidic Extracts of Florida Phosphate Mining Materials Using Chelating Polymer 1-Octadecene, Polymer with 2, 5-Furandione, Sodium Salt. *Minerals* **2019**, *9*, 477. [CrossRef]
8. Loutou, M.; Misrar, W.; Koudad, M.; Mansori, M.; Grase, L.; Favotto, C.; Taha, Y.; Hakkou, R. Phosphate Mine Tailing Recycling in Membrane Filter Manufacturing: Microstructure and Filtration Suitability. *Minerals* **2019**, *9*, 318. [CrossRef]
9. Amrani, M.; Taha, Y.; Kchikach, A.; Benzaazoua, M.; Hakkou, R. Valorization of Phosphate Mine Waste Rocks as Materials for Road Construction. *Minerals* **2019**, *9*, 237. [CrossRef]
10. Liu, Y.; Zhang, Q.; Chen, Q.; Qi, C.; Su, Z.; Huang, Z. Utilisation of water-washing pre-treated phosphogypsum for cemented paste backfill. *Minerals* **2019**, *9*, 175. [CrossRef]
11. Li, X.; Zhou, S.; Zhou, Y.; Min, C.; Cao, Z.; Du, J.; Luo, L.; Shi, Y. Durability Evaluation of Phosphogypsum-Based Cemented Backfill Through Drying-Wetting Cycles. *Minerals* **2019**, *9*, 321. [CrossRef]
12. Shen, Z.; Zhang, Q.; Cheng, W.; Chen, Q. Radioactivity of Five Typical General Industrial Solid Wastes and Its Influence in Solid Waste Recycling. *Minerals* **2019**, *9*, 168. [CrossRef]

© 2019 by the authors. Licensee MDPI, Basel, Switzerland. This article is an open access article distributed under the terms and conditions of the Creative Commons Attribution (CC BY) license (http://creativecommons.org/licenses/by/4.0/).

Review

Review on Beneficiation Techniques and Reagents Used for Phosphate Ores

Yaoyang Ruan, Dongsheng He and Ruan Chi *

Xingfa School of Mining Engineering, Wuhan Institute of Technology, Wuhan 430074, China; ruanyaoyang@163.com (Y.R.); csuhy@126.com (D.H.)
* Correspondence: rac@wit.edu.cn

Received: 30 January 2019; Accepted: 24 April 2019; Published: 25 April 2019

Abstract: Phosphate ore is an important raw material for manufacturing fertilizers and phosphorous chemical products. While most of the phosphate resources cannot be directly treated as feed stock due to the low grade of P_2O_5 and high content of impurities. In order to obtain a qualified phosphate concentrate, the beneficiation of the low-grade phosphate ore is, hence, of great necessity. Many beneficiation techniques can be employed to upgrade the P_2O_5 grade of phosphate ores based on their characteristics in chemical composition and texture. The flotation process is most widely applied to balance the P_2O_5 recovery ratio and cost. In this review, the dominant techniques for the beneficiation of phosphate ores are introduced. Moreover, the factors that affect the flotation of phosphate ore, including the properties of mineralogy, flotation reagents (depressants and collectors) and flotation medium, were systematically analyzed.

Keywords: phosphate ore; beneficiation; mineralogy; depressant; collector; interfering ions

1. Introduction

Phosphate ore is an essential raw material for manufacturing phosphoric industrial products, and it is irreplaceable. It has been widely utilizing in agriculture, chemical industry, food, pharmacy, etc. The world's phosphate reserves total up to 70 billion tons. Morocco has the biggest phosphate resources, a reserve of 50 billion tons, accounting for 71.43% of the total amount [1]. China, Morocco, the United States and Russia are the leading countries of phosphate production with a proportion of 79%. Other countries including Brazil, Jordan, Egypt and Saudi Arabia take up the rest of the production. There are four major types of phosphate resources according to the mineralization, viz., igneous deposits, metamorphic deposits, sedimentary deposits and biogenic deposits (guano accumulations). Approximately 75% of phosphate resources is attributed to sedimentary origin [2].

The high-grade phosphate ore will go through the wet process and pyrogenic process respectively to obtain the intermediate products of phosphoric acid and phosphorus, which can be used to produce various phosphate fertilizers and phosphates. The requirements for phosphate concentrate used in wet phosphoric process are (1) a P_2O_5 grade higher than 30%, (2) a CaO/P_2O_5 ratio less than 1.6, and (3) a MgO content less than 1% [3]. However, with an increasing population and a demand for phosphate, the high-grade phosphate ores with a low content of impurities are being depleted. Most of phosphate ores are not suitable for direct use in the acidulation process because they have a relatively low P_2O_5 content and generally contain a series of gangue minerals, mainly quartz, mica, feldspar, dolomite, calcite, clays and so on. Therefore, the industry of phosphate beneficiation is confronted with a great challenge, i.e., how to exploit these low-grade phosphate ores in economical and efficient way [4].

Since the low-grade phosphate ores must be pretreated to reach a qualified phosphate concentrate, the beneficiation techniques and reagents become extremely critical to achieve the selective separation of phosphate minerals. Based on the diverse compositions and texture of the run-of-mine, the corresponding beneficiation processes and reagents are introduced. This paper comprehensively reviews the phosphate beneficiation techniques, reagents and the factors affecting phosphate flotation, aiming at providing a guideline for the development and utilization of phosphate ores in the future.

2. Beneficiation Techniques for Phosphate Ores

2.1. Flotation

Due to the high efficiency in the removal of silicate and carbonate gangue minerals, froth flotation is dominantly employed for the beneficiation of phosphate ore, especially for low-grade sedimentary phosphate rocks. Generally, direct flotation by anionic surfactants or reverse flotation by cationic surfactants are conducted to remove the silicate gangues from phosphate ores. During the direct flotation process, the pulp pH was adjusted to around 9.5 by adding soda ash, and a proper amount of water glass was required simultaneously to depress the float of silicate minerals. Then, the phosphate minerals turned out to be floated with the assistance of anionic surfactants. For the reverse flotation process, silica was floated using cationic surfactants in a weak acidic or neutral pH, and the phosphate minerals (dominantly apatite) were obtained from the underflow product. The superior adsorption of cationic surfactants on silica rather than apatite probably can be attributed to the more negatively charged surfaces of silica [5]. The conventional "Crago" double float process is used to remove silica twice from phosphate ore in Florida. Due to the higher fatty acid prices, lower feed grade and stricter environmental regulations, a new reverse "Crago" process comprised of amine-fatty acid flotation was developed [6], and its schematic flowsheet is given in Figure 1. For the removal of carbonate gangues, a high separation efficiency can be achieved using reverse anionic flotation at around pH 4.5 without the addition of a phosphate depressant [7].

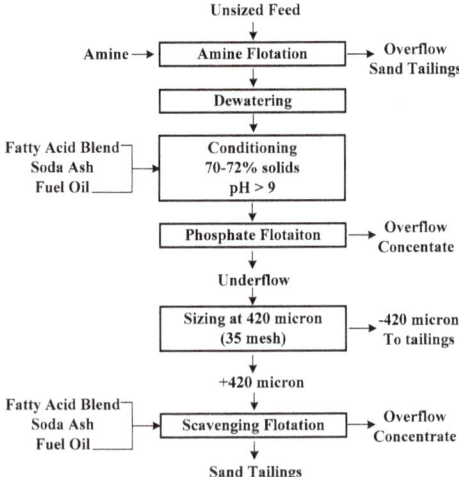

Figure 1. The reverse "Crago" double float process for beneficiation of Florida phosphate ore [6].

The phosphate ores mainly contain the siliceous phosphate rock, calcareous phosphate rock and calcareous-siliceous phosphate rock. Different types of phosphate ores should be subjected to specific flotation processes for upgrading according to the mineral composition characteristics. Both of the anionic direct flotation and cationic reverse flotation are applied for the beneficiation of siliceous phosphate rock, while the single anionic reverse flotation is preferred to beneficiate the calcareous phosphate rock. A two-step flotation process, named direct-reverse flotation (Figure 2) or double reverse flotation, is usually carried out to remove silica and carbonate gangue minerals stepwise from the calcareous-siliceous phosphate rocks [8].

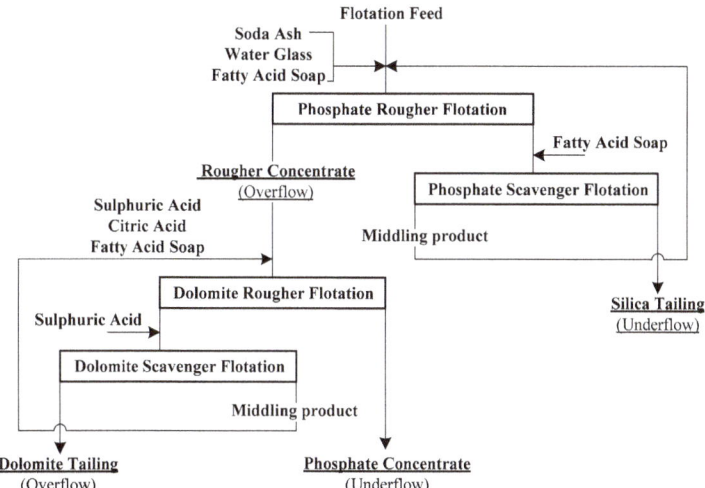

Figure 2. The direct-reverse flotation process for beneficiation of calcareous-siliceous phosphate ore [9].

However, it is found that the reagent consumption in direct flotation mid-low grade phosphate rocks is relatively higher and, thus, increases operating cost. At the same time, the recycled water is also difficult to reuse, and the high temperature is required sometimes, which results in the application of the direct process being highly restricted. Thereof, it is only suitable to beneficiate the low-grade phosphate ores. At present, the single reverse flotation is attracting concentrated focus and has been widely applied to beneficiate phosphate ores in China.

In some cases, an appropriate amount of silica in the phosphate concentrate is desired for the production of phosphoric acid by the wet process. As we know, the fluorine in phosphate ores will react with acid to generate HF and further corrode the surrounding facilities. However, this adverse effect can be greatly reduced, since the silica reserved in the flotation concentrate can react with HF to form SiF_4 and fluorosilicates [10]. The production practice has shown that it is economically feasible to obtain a qualified phosphate concentrate using single reverse flotation in case of the high-magnesium calcareous phosphate rock is mixed uniformly with a certain proportion of siliceous phosphate rock.

2.2. Attrition Scrubbing and Desliming

This technique is commonly used for the treatment of weathered phosphate ores that contains a relatively high content of clays. In this case, a coarse phosphate concentrate can be obtained after the elimination of fine-grained clays. This method has been successfully applied on the high-grade weathered phosphate ores in Dianchi area (Yunnan, China). As the weathered phosphate ores were crushed and sieved, the fraction of −25 mm was used to conduct the scrubbing and desliming; as a consequence, a high-quality concentrate with a grade of P_2O_5 above 32% was obtained after the gangue clays that dominant in sesquioxide ($Al_2O_3 + Fe_2O_3$) were removed [11,12]. At the same time, this process has also been applied in the Crago flotation process for the preliminary enrichment of phosphorites ahead of the anionic-cationic flotation section [6,13].

However, it should be noted that there is still a certain amount of phosphorus remaining in the scrubbed tailings, and its grade is nearly up to 19%. If phosphorus slurry deposits into the tailings pond instead of being effectively recovered, it will cause a big waste of valuable resources and a nonnegligible environmental issue. He et al. confirmed that using a direct flotation can effectively recover the phosphorus from the scrubbed tailings in Yunnan and finally obtained a concentrate assaying 28.26% P_2O_5 with a recovery of 80.37% [12].

2.3. Electrostatic Separation

Electrostatic separation is always used to pre-concentrate grained phosphate minerals by discarding a certain proportion of silica. This approach can reduce the energy consumption of grinding and the cost of chemical reagents and water usage. Sobhy and Tao investigated the feasibility of using rotary triboelectrostatic separator in a dry beneficiation process of Florida phosphate ores [14]. Their results showed that a concentrate containing 30% P_2O_5 with a recovery of more than 85% was collected and that the acid insoluble rejection also reached almost 90%. The Saudi phosphate ores are always in grain size with 9–70 mm and can be upgraded by an electrostatic separator (COM Tertiary XRT B2400, GREANEX, Wilmington, DE, USA). In the laboratory experiment and industrial test, respectively, 60.2% and 44.65% of silica was demonstrated to be removed [15]. However, this method is usually restricted by the low capacity of the electrostatic separators and, thus, has not been applied for the large-scale production of phosphate concentrate.

2.4. Magnetic Separation

Magnetic separation is successfully applied for the beneficiation of igneous apatite deposits, which is abundant in ferriferous minerals, such as magnetite or titanomagnetite. The beneficiation of apatite from the "Barreiro" complex carbonatite phosphate ore in Brazil can be realized by a consecutive low-intensity magnetic separation, desliming, flotation and high intensity magnetic separation process [16]. Meanwhile, it is confirmed that magnetic separation has a potential application on the treatment of coarse-grained sedimentary phosphate ore for the removal of magnetic gangues. Blazy and Jdid conducted a high-gradient magnetic separation (HGMS) on the Egyptian Abu Tartur high-grade phosphate ore in the size range of 38–210 mm. A concentrate assaying 31.2% P_2O_5 with a recovery of 70% was obtained through the removal of dolomite gangue minerals [17]. It was observed that the ferriferous dolomite selectively entered into the magnetic product during the magnetic separation process, while the phosphate minerals remained in the nonmagnetic part. Another research also demonstrated that the removal of dolomite carbonates from a marine sedimentary phosphate ore (80–250 µm) can be achieved through magnetic the separation process [18].

The exploration of the magnetic separation process on the fine-grained rock phosphate ores was undertaken by Shaikh and Dixit [19], and the schematic flowsheet was given in Figure 3. Phosphate ores (53–63 µm) from Rajasthan and Madhya Pradesh were subjected to a two-stage high-gradient magnetic separation. In the first stage, the calcite/iron was coated by magnetite and effectively separated as the magnetic product under the effect of sodium oleate and dipotassium hydrogen phosphate. The phosphorite and silica remained in the nonmagnetic proportion. In the second stage, the phosphorite was selectively coated by the magnetite again and consequently enriched in the final magnetic product in the presence of sodium oleate and sodium metasilicate. A final concentrate containing about 31.5% P_2O_5 and 8.8% SiO_2 was obtained under the optimized condition while the overall recovery of P_2O_5 was only 65.0%. It can be seen from the above research that the phosphate recovery is low during the magnetic separation, and thus, this method is not efficient for the beneficiation of fine-grained phosphate ores.

2.5. Gravity Separation

Gravity separation is carried out according to the density difference between valuable minerals and gangue minerals. Because of the similar density of apatite and gangue minerals, it is usually difficult to recover phosphate minerals that are disseminated into fine grains. However, the coarse phosphate minerals with a strip structure in the sedimentary phosphate rock were proved to be effectively beneficiated by discarding tailings through gravity separation. In this case, the energy consumption for grinding is drastically reduced. The band-shaped phosphate rock in the Yichang (China) area contains an abundance of phosphate minerals which are mainly distributed ranging from

2 mm to 18 mm. It is feasible to adopt gravity separation to beneficiate this type of phosphate rock. The typical flowsheet of gravity separation for phosphate rocks is given in Figure 4.

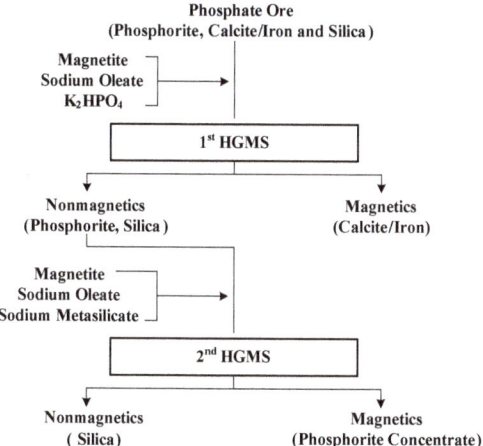

Figure 3. Schematic flowsheet of high gradient magnetic separation for Rajasthan and Madhya Pradesh rock phosphate [19].

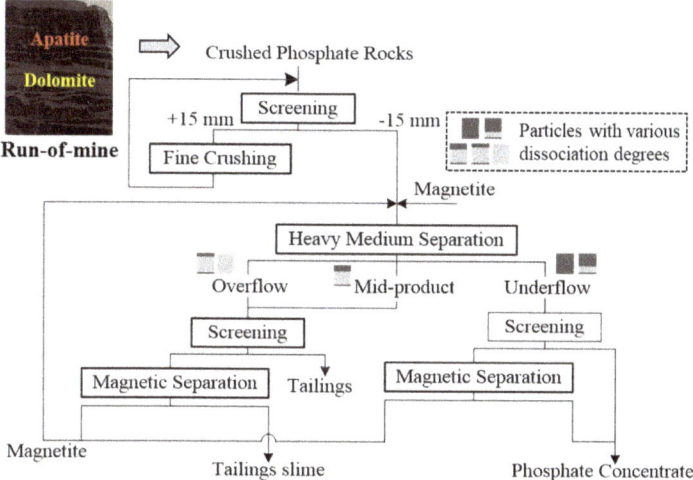

Figure 4. Schematic flowsheet of gravity separation for sedimentary phosphate rock characterized by a strip structure.

Through the addition of magnetite (nearly 92% passing 0.1 mm), the density of the pulp can be adjusted to 2.85–2.91 g/mL, which further facilitates the separation of the valuable and gangue minerals. After subjecting the pulp to gravity separation, the desired minerals are collected in the underflow through sedimentation while the gangue minerals report to the mid-product and overflow. This is often achieved by using a heavy medium cyclone. The mid-product and overflow product are usually merged in order to sieve out the coarse-sized tailings which can be used for mine backfilling. Magnetite contained in the underflow and overflow can be effectively recycled through a medium draining screen followed by magnetic separation. Afterwards, the nonmagnetic materials will be further subjected to concentration and filtration, and a fine-grained concentrate and tailing slime can be obtained. This method has been successfully applied on the Yichang phosphate ore which contains 23.50% P_2O_5. The content and recovery of P_2O_5 in the concentrate are up to 30.64% and 85.60% respectively. The annual capacity of the local plant has enlarged to 1.2 million tons since the year of 2008 in the Yichang area [20]. As an economical and environmentally friendly beneficiation technique, gravity separation is increasingly popularized and applied by some phosphate mining enterprises in China.

So far, the physical beneficiation processes such as scrubbing, gravity separation and electrostatic separation processes have not achieved large scale application because it is still difficult to obtain satisfactory phosphate concentrates. At the same time, the P_2O_5 content in the tailings is still high, which results in a low P_2O_5 recovery from these processes. In order to obtain qualified concentrates and to maximize mineral utilization simultaneously, it is necessary to combine these processes with flotation to form a joint beneficiation process [21]. At present, the main combining processes which have been scaled up include the scrubbing-flotation process, the magnetic-flotation process and the gravity-flotation process.

2.6. Calcination

More than 10% of marketable phosphates are produced by calcination in the world [2]. During the calcination process, calcareous minerals are decomposed into calcium oxide and magnesium oxide at temperature about 950 °C, and carbon dioxide is released. Then, the calcined phosphate rocks are subjected to quenching by using NH_4Cl or NH_4NO_3 solution at a concentration of about 5%. Subsequently, the fine grained $Ca(OH)_2$ and $Mg(OH)_2$ are discarded as the slime phase after a classification by hydrocyclone, and the coarser high grade concentrate are successfully separated. For those phosphate ores containing organic matter, it is also an efficient technique for upgrading at a temperature of 650–750 °C [22].

However, it consumed large amounts of energy during the calcination, making this process only suitable for Mideast countries that have low energy cost and limited water sources. Compared to the total energy required for the physical beneficiation of phosphate ores, the consumption needed for calcination is doubled [2]. Furthermore, the solubility and reactivity of the calcined phosphate rocks may decrease during the manufacture of phosphoric acid by the wet process. Watti et al. [23] pointed out that the solubility of a calcined product sharply decreased with the increase of temperature because the dissolved P_2O_5 in a solution of citric acid varied from 9.32% to 3.55% when the treatment temperature increased from 25 °C to 1000 °C. It also has been revealed that the quenching process for the calcined ore was inefficient with the fine sizes and at low temperatures [24]. The above results showed that there are still some technical challenges remaining in the calcination of calcareous phosphate ore.

2.7. Acid Leaching

The calcareous phosphate ores can be also upgraded by acid leaching. In this process, the carbonate gangue minerals prefer to dissolve in the dilute acid solution, while the phosphate minerals remain in the leached residues. As a result, the separation of the phosphate minerals against carbonate gangue minerals is achieved. Using common mineral acids often results in the dissolution of P_2O_5 and causes the loss of P_2O_5 recovery while the leaching selectivity of the organic acids proves high. Therefore, the organic acids are preferred in the leaching process. The commonly used organic acids are lactic acid [25], formic acid [26] and succinic acid [27].

In order to obtain a satisfactory removal of the carbonate gangue minerals, it is necessary to optimize the leaching parameters such as the acid concentration, reaction time, particle size, solid to liquid ratio and reaction temperature [28]. Gharabaghi et al. [29] found that the leaching process for a low-grade calcareous phosphate rock should be optimized to obtain the qualified concentrate when using acetic acid. The best result was achieved by using 15 wt. % of acetic acid concentration at a liquid to solid ratio of 15:1, an optimum reaction time of 60 min and a leaching temperature at 40 °C. After leaching, the P_2O_5 content can be improved to 32.1% with the P_2O_5 recovery of 81%. Although the chemical leaching process of the calcareous phosphate rock is relatively simple and the qualified concentrate can be obtained under the optimized leaching conditions, there are still some drawbacks, such as (1) the involved organic acids are costly, which greatly increases the production cost as well, and (2) the leaching process usually requires 30–60 min [30], which is much longer than other physical beneficiation methods. Abu-Eishah et al. [28] found that the demanded leaching time was more than 1.2 h to deal with the phosphate ores. The long leaching time greatly limits the capacity of plants and restricts the large-scale production of phosphates. As a result, the profit of plant has been reduced.

From the discussion above, beneficiation techniques and their applications for various phosphate ores are summarized in Table 1.

Table 1. Beneficiation techniques and their applications for various phosphate ores.

Beneficiation Techniques	Phosphate Rock Types	Applications
Flotation	Siliceous phosphate rock Calcareous phosphate rock Calcareous-siliceous phosphate rock	Widely applied in the beneficiation of various phosphate rocks on a large scale, especially for refractory sedimentary phosphate ore
Attrition scrubbing-desliming	Weathered phosphate ores	Used for discarding clay minerals and eliminating the detrimental effect of slimes on the subsequent flotation process
Gravity separation	Sedimentary phosphate rock characterized by a strip texture	Discarding gangues to achieve the preconcentration of phosphate minerals
Magnetic separation	Phosphate ores containing magnetic gangues	
Calcination	Calcareous phosphate rock	Acceptable for areas that have low cost energy and limited water resources
Electrostatic separation	Coarse gained siliceous phosphate rock	Mostly are experimental studies in laboratory scale
Acid leaching	Calcareous phosphate rock	

3. Factors Affecting the Flotation of Phosphate Ore

Satisfactory separation results can be easily obtained by a relatively simple flotation technique when beneficiating phosphate ore with a high degree of crystallinity. However, for the sedimentary phosphate ore characterized by fine grained dissemination and complex chemical composition, it is very difficult to obtain a high-grade phosphate concentrate with an acceptable recovery rate. In addition, as flotation is a gas-solid-liquid three-phase interface reaction process, there are many factors that can affect phosphate ore flotation, such as, mineralogical properties, flotation reagent properties and flotation medium properties.

3.1. Mineralogical Properties

3.1.1. Mineral Type

The influence of crystal chemistry properties on the floatability of apatites was systematically investigated by Rodrigues and Brandao [31], and the curves was illustrated in Figure 5. They stated that the floatability of different types of apatite decreased in the order of pegmatitic apatite (Durango, C. Grosso and Camisao), igneous apatite (Tapira and Catalao), metamorphic apatite (Monteiro), metasedimentary apatite (Itataia) and sedimentary apatite (Igarassu). The floatability of pegmatitic apatites is excellent with a value approaching 100% at a wide pH range. The maximum floatability was observed at neutral pH for the floatability of igneous and metamorphic apatites, and the floatability of metamorphic apatites decreased faster than that of igneous apatites under acid and alkaline conditions. Generally, the floatability of the metasedimentary apatite and sedimentary apatite is poor and the recovery is low in the whole variation range of pH value.

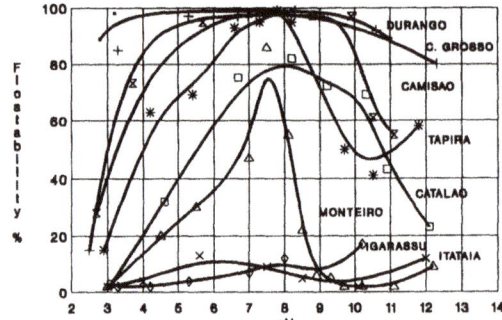

Figure 5. Microflotation curves of various apatite samples using 2.5 × 10^{-5} mol/L sodium oleate [31].

Through the determination of the unit cell parameters, it was found by Rodrigues and Brandao that the unit cell parameter of igneous, metamorphic and sedimentary-metamorphic apatite is higher than that of sedimentary apatite. The degree of crystallinity decreased in the rank regarding the geological origin: pneumatolitic-hydrothermal, pegmatitic, igneous, metasomatic, metasedimentary and sedimentary. It can be concluded that there is a positive correlation between the floatability and crystallization degree of apatite. They identified that apatites with a higher degree of crystallinity displayed much better floatability, while apatites with a lower crystallinity level needed a higher collector concentration to reach the same floatability corresponding to a highly crystalline one. The worse flotation performances of the apatites having lower crystallinity levels is believed to relate to their higher solubilities, and less stable mineral/water interfaces [31].

3.1.2. Mineral Granularity

It is known that the particle size distribution of the flotation feed has a significant effect on the separation efficiency. Generally, the processibility of coarse-grained minerals is better than that of the fine-grained ones. It is probably attributed to the high surface energy of the fine size, contributing to an increased dissolution from the surface of particles. Thus, the nonselective adsorption of the reagents and undesirable entrainment or entrapment of fine particles occurred [32]. Moreover, the commonly used collectors in phosphate plant, amine and fatty acids are sensitive to the minerals with a fine particle size. How to realize an efficient separation of fine particles is an urgent technical problem faced by the phosphate beneficiation plant. Recently, the effect of particle size on the flotation performance of phosphate ore has been reported.

The flotation selectivity of apatite from a phosphate ore as a function of particle size has been investigated by Santana et al. [32]. They found that the P_2O_5/Fe_2O_3 and P_2O_5/SiO_2 ratios of coarse fractions (+65, +100 and +150 mesh) were higher than the industrial requirements but that there were no satisfied selectivity observed when using finer fraction (−400 mesh). Another study reported by Santana showed that a phosphate concentrate with a high grade and satisfactory recovery were not feasible from both of the coarser (+65 and +100 mesh) and finer fractions (−400 mesh). The optimum particle size range for apatite flotation is from 37 μm to 105 μm [33]. Cationic flotation test conducted by Guo and Li showed that the P_2O_5 grade and recovery of 45–74 μm fraction are higher than those of −45 μm fraction about 2% and 5%, respectively [34].

Mineral liberation is closely related to the particle size in the flotation process of actual phosphate ores. However, there is a certain conflict between liberation and particle size for the beneficiation of fine-grained phosphate rocks because fine grinding is desired to achieve a higher liberation degree and the obtained fine particles will adversely affect the separation efficiency. On the contrary, the separation efficiency of coarse particle without adequate liberation is limited as well [35]. Therefore, the effective liberation of desired minerals as well as avoid overgrinding of ores is of great importance for the subsequent flotation process.

3.2. Properties of Flotation Reagents

A flotation reagent, especially depressants and collectors, has a great impact on the selective separation of phosphate minerals and gangues. During the flotation process, the surface of specific mineral becomes hydrophilic due to the hydrophilic film formed after the adsorption of depressant. Meanwhile, collector molecules are oriented on the surface of the floated mineral, causing the enhancement of hydrophobicity. For example, the addition of depressant for gangue mineral in the direct flotation of phosphate ore is often required. After the depression of gangues, the phosphate minerals can effectively react with collectors and become more gathered by attaching to the rising bubbles.

3.2.1. Depressants

(1) Depressants of Silicate Minerals

Sodium silicate is an effective depressant in the direct flotation of apatite from siliceous phosphate ore using fatty acids [36]. Compared to the infrared spectra of quartz in the absence and presence of sodium silicate at pH 7, the adsorption of dominant hydrolysis species $Si(OH)_4$ was detected by Silva et al. [37]. The quartz depression mechanism was elucidated through the ligand exchange model, where pairs of electrons were shared by sites of the surface and species of silicate. Generally, the modulus and dosage of sodium silicate have a great impact on the depression of quartz occurred in flotation process [38].

The depression of quartz caused by adding $NaHCO_3$ in the pulp was also observed in both the cationic flotation using amine and the anionic flotation using sodium oleate and Ca^{2+} as activators. Sayilgan and Arol thought that the depression might result from the compression of the double layer or competitive adsorption of Na^+ and amine ions on the surface of quartz in the case of cationic flotation. For the anionic flotation, it was convincible that the activator calcium ions could react with the carbonate ions and precipitate, making the available concentration of $Ca(OH)^+$ for quartz activation apparently reduced [39].

Starch exhibited a depression effect on the flotation of quartz. Pavlovic and Brandao identified that the adsorption of starch on the surface of quartz should be attributed to its flocculation property [40]. However, starch was not an efficient depressant for quartz due to the similar depression might have occurred in the flotation of apatite and dolomite. Considering the cost rather than the quality of the product, starch is still widely applied in the flotation of igneous phosphate ores in Brazilian concentrators [41].

(2) Depressants of Carbonate Minerals

In the phosphate beneficiation practice, carbonate gangues can be efficiently removed by a reverse flotation process with the addition of a phosphate depressant. However, for the upgrading of siliceous calcareous phosphate ore, two types of gangue minerals need to be removed by a double float process. In order to simplify the flotation process, attempts of the concurrent depression of silicate and carbonate gangues should be undertaken to achieve the phosphate enrichment at one stage. Therefore, it is of great importance to investigate the depressant for carbonate minerals. Zheng and Smith evaluated different organic chemical reagents which functioned as dolomite depressants in the flotation of single mineral and mixed apatite and dolomite, respectively. The results indicated that carboxymethyl cellulose, citric acid and naphtyl anthyl sulfonates were effective depressants for dolomite [42].

The selective separation of collophane and dolomite using β-naphthyl sulfonate formaldehyde condensate (NSFC) as a depressant by anionic flotation at pH 9 was conducted by Yu et al. [43]. More significant changes were observed in the case of dolomite, and chemical adsorption occurred on the dolomite while a weak adsorption on collophane through hydrogen bonds was identified. Although the high efficiency of dolomite depression can be achieved by NSFC, there is a critical obstacle for the large-scale application due to its chemical toxicity.

The depression effect caused by the adhesion of bacteria, namely *Bacillus subtilis* and *Mycobacterium phlei* on the anionic flotation of apatite and dolomite was elucidated by Zhang et al. [44]. It was found that these two bacteria adsorbed on the surface of dolomite and functioned as depressants for both of dolomite and apatite. It seems to be difficult for the separation of apatite from dolomite using these bacteria as depressant, but they can modulate the flotation environment and provide a new idea for the development of depressant for dolomite and apatite.

(3) Depressants of Phosphate Minerals

The depression of apatite can be achieved by adjusting the pH of pulp because the flotation behavior barely occurred at pH below 4.5 in an anionic flotation of apatite without the addition of any other phosphate depressant [5]. It is evident that the species of Ca^{2+}, $CaH_2PO_4^+$ and $H_2PO_4^-$ are dominant at pH below 4.5 and that their concentration increases as the decrease of pH according to the solution equilibria of dissolved apatite [45]. It can be deduced that the species of $CaH_2PO_4^+$ and $H_2PO_4^-$ should be responsible for apatite depression. $H_2PO_4^-$ prefers to bond with calcium ions exposed on the surface of apatite, resulting in the reduction of available active sites [44]. In that case, inorganic acids and various phosphate salts, such as potassium dihydrogen phosphate and sodium pyrophosphate, can be used as an efficient depressant for apatite [46,47].

In addition, it was reported that the addition of sulfate or oxalate salts can enhance the depression of apatite in acid media [48,49]. El-Mofty and El-Midany claimed that the dissolved Ca^{2+} can precipitate in the presence of oxalate or sulfate ions, causing more dissolution of apatite, and thus, more phosphate ions existed in aqueous solution [49]. Though sulfuric acid and its salts demonstrate a relatively strong depression on the phosphate minerals, it is easy to cause pipe fouling in production due to the formation of gypsum. Consequently, phosphoric acid is considered for application in the anionic reverse flotation for a beneficiation plant.

Aiming at enhancing the P_2O_5 recovery in cationic reverse flotation, various depressants including sodium tripolyphosphate, fluosilicic acid, diphosphonic acid and starch were evaluated by Zhang and Snow [50]. It was found that starch was an ideal phosphate depressant in the removal of fine (−35 mesh) silica from phosphate, while sodium tripolyphosphate may be the best one for the beneficiation of coarser (+35 mesh) feeds. The phosphate depression caused by a low molecular weight polyacrylamide containing both hydroxyl and carboxylic acid functional groups was also investigated by Nagaraj et al. [51]. The polymer demonstrated an excellent selectivity in the separation of apatite from siliceous gangue using reverse cationic flotation. Various depressants used in the flotation of phosphate ores are collected and presented in Table 2.

Table 2. Depressants used in the flotation of phosphate ores.

Mineral to Be Depressed	Name of Depressant	References
Silicate minerals	Sodium silicate	[36,52]
	Sodium and calcium lignin sulfonates	[53]
	Copolymers or terpolymers derived from acrylamide units and N-acrylamidoglycolic acid units	[54]
Carbonate minerals	Carboxymethyl cellulose, citric acid, naphtyl anthyl sulfonates	[42]
	β-naphthyl sulfonate formaldehyde condensate	[43]
Phosphate minerals	Potassium dihydrogen phosphate, sodium pyrophosphate	[46,47]
	Sodium tripolyphosphate, fluosilicic acid, diphosphonic acid, starch	[50]
	Cashew gum	[55]
	Sulfuric acid, phosphoric acid	[46,56]
	Alkyl phosphate acids, hydrofluoric acid	[57,58]
	Dipotassium hydrogen phosphate	[59]
	Cellulase enzyme	[60]
	Iron/aluminum sulfate, tartaric acid	[61,62]

3.2.2. Collectors

(1) Anionic Collectors

Long chain fatty acids are commonly used anionic collectors in phosphate flotation, especially for sodium oleate. With the aid of atomic force microscopy, the adsorption behavior of potassium oleate and the precipitation of Ca^{2+} ions on the surface of the apatite crystal were visualized. It is believed that the formation and precipitation of calcium dioleate occurred in the form of agglomerates on the apatite surface. The presence of agglomerates made the apatite surface rougher and heterogeneous, which contributing to the flotation behavior of apatite [63]. In the past decades, tall oil and oxidized petroleum were mainly used as the source of manufacturing anionic collectors for phosphate flotation. Due to the high temperature requirement, low flotation selectivity and relatively huge reagent consumption, alternative sources for fatty acids can be derived from the less expensive vegetable oil, such as rice bran oil, hydrogenated soybean oil, cottonseed oil and jojoba oil [9,41,64].

To further improve the solubility and selectivity of fatty acid soap at an ambient temperature, the modified fatty acids characterized by a multifunction group was designed and synthesized though chlorination, hydroxylation and etherification. The flotation results of a low-grade collophanite showed that the modified soybean oil had a better performance than a conventional fatty acid collector with a less dosage [65]. For those anionic surfactants with a low foamability, such as sodium dodecyl sulfonate and sulphosuccinate, they are mostly used as an auxiliary reagent to improve the flotation performance of reagent in practice [66].

(2) Cationic Collectors

In the practical flotation of phosphate ores, cationic collectors are dominantly used for the removal of silicate minerals. Guo and Li claimed that the alkyl amine salt (DAH) is a more effective collector for removing the silica from the fine siliceous calcareous phosphate ore as compared to the ether-amine salt or the quaternary ammonium salt (CTAB) [67]. The flotation behavior of pure quartz studied by Sahoo et al. indicated that cetyl pyridinium bromide and benzyl dimethyl tetradecyl ammonium chloride were more effective than the conventional cationic collector DDA and CTAB [68]. Flotigam EDA 3 (R–O–CH_2–CH_2–C–NH_2) and Flotigan 2835-2L (R–O–$(CH_2)_3$–NH–$(CH_2)_3$–NH_2), manufactured by Clariant, were selected as the collector for quartz flotation. It was observed that ether diamine was more effective in the flotation of medium and coarse quartz, while ether monoamine performed better in the case of fine fraction [69]. On the other hand, the above cationic collectors were used for the dephosphorization of an iron ore. Flotigam EDA at a dosage of 150 g/t led to a 0.201% phosphorus concentrate and mass recovery of 62.31%, while the use of Flotigam 2835-2L presented a phosphorous content of 0.31% and mass recovery of 90.24% [70].

The advantage of the cationic collector is that it can achieve the separation of valuable minerals from gangues at a low temperature with less consumption of the reagent. However, amine collectors are generally sensitive to the slimes contained in the run-of-mine ore. Additionally, attributed to the high foam viscosity, it is difficult to defoam during the flotation process when using an amine as the collector. Therefore, cationic collectors are suitable for the flotation of coarse-grained ores, but they show relatively poor performance in the separation of fine-grained materials.

(3) Amphoteric Collectors

This type of collector can be used for the removal of dolomite from phosphate ores. Their flotation selectivity is relatively high over a wide pH range and negligibly influenced by the dissolved species and temperature. It is reported that a phosphate concentrate assaying above 32% of P_2O_5 with a recovery around 90% can be obtained from a mixture of pure phosphate and dolomite minerals when using dodecyl-N-carboxyethyl-N-hydroxyethyl-imidazoline as the collector [71,72]. The separation of dolomite from Abu-Tartur phosphate ore using a similar amphoteric collector was conducted by Elmahdy et al. [73]. Under the determined optimal flotation parameters, a concentrate of 0.53% MgO and 31% P_2O_5 with a recovery of 90% was obtained. Measurements of the zeta potential and adsorption in the absence and presence of amphoteric surfactant were performed to elucidate the mechanism. The results indicated that a much greater variation was observed on the calcite surface when compared to the case of apatite surface at an alkaline pH [74,75]. Therefore, the amphoteric surfactant can preferentially be adsorbed on the surface of calcite and can achieve the selective separation of calcite and apatite. However, most of the amphoteric collectors are limited to laboratory studies, and the real industrial application has been barely reported.

(4) Mixed Collectors

The mixture of various surfactants is the main research issue of the development of collector in the flotation of phosphate ore, especially for an ionic and nonionic surfactant mixture. Some mixed collectors used in phosphate flotation is presented in Table 3. It is well-known that mixed collectors are supposed to have a number of synergistic advantages over an individual surfactant. The presence of nonionic surfactants can enhance the adsorption of ionic surfactants on the surface of minerals due to the hydrophobic chain interactions and the reduction of electrostatic repulsion between ionic head groups [76,77]. The measurement of the contact angle of sodium oleate on the apatite surface in the absence and presence of the ethoxylated nonylphenol surfactant (NP-4) showed that the beneficial effect of the nonionic surfactant was achieved at low oleate concentrations (0.1–5 mg/L). It was further confirmed that the activity of ionic surfactants is enhanced with the addition of nonionic surfactant due to the synergistic effect in the formation of mixed micelles and the reduction of surface tension [78,79]. However, it should be noted that the excessive addition of the nonionic surfactant would cause an adverse effect on the flotation process. A decreased adsorption of reagent on apatite surface was found by Sis and Chander when using the mixture of sodium oleate and NP-4 at a weight ratio of 2:1 [80]. They thought that the adsorption decrease caused by nonionic surfactant could be attributed to the competition between the surfactants or the prevention of oleate ions from the precipitation as calcium oleate salt [80,81].

Mixing the anionic collector with nonionic surfactants is also desirable in improving the selectivity of the flotation process. It is reported that the alkyl-hydroxamic acid had a higher selectivity in phosphate flotation than the traditional fatty acid/fuel oil. When it was further used as a collector in the flotation of francolite together with alcohol, the francolite recovery was drastically enhanced [8]. Base on the high-speed video images which clearly exhibited the detachment of collector mixture drop from the quartz surface and then spread on apatite surface, the selective attachment of hydroxamic acid and alcohol collector mixtures in phosphate flotation was confirmed [82].

The effect of mixing anionic collectors with nonionic surfactants on the selective separation of Ca-bearing minerals, including calcite and francolite, has been investigated by Filippov et al. [76]. A synergetic behavior of mixed collectors of anionic surfactants such as alkyl hydroxamate (AERO 6493), di-2-(ethylhexyl) phosphoric acid (D2EHPA) and iso-tridecanol (PX4826) has been shown in the flotation data on francolite and calcite. It is noted that the difference in floatability of francolite and calcite was significantly enhanced at pH 8 when using a mixture of AERO 6493 and PX4826 without the presence of calcite depressant. Furthermore, synergistic effects of primary amine and PX4826 in terms of calcite and apatite recoveries by flotation at pH 8 were also observed [83]. The dephosphorization of magnetite ore by froth flotation showed that the anionic flotation response of apatite increased in the presence of nonionic ethaoxylated nonylphenol with no effect on the magnetite flotation [84].

Table 3. Mixed collectors used in the flotation of phosphate ores.

Type of Mixed Collector	Primary Collector	Auxiliary Reagent	References
Anionic-anionic	Fatty acids	petroleum sulfonates and ethoxylated alcohol ether sulfates	[85]
	Fatty acids	alkyl aryl sulphonate and sulphated fatty acid	[86]
	Oleic acid	Sodium dodecyl benzene sulfonate and oxidized paraffin soap	[87]
Cationic-anionic	N-aminoethylpiperazine	Fatty acids	[88]
Cationic-nonionic	Primary amine (Cataflot)	Iso-tridecanol (PX4826)	[83]
Anionic-nonionic	Fatty acids	hydrocarbon oil such as kerosene or fuel oils	[89]
	Fatty acids	Esters of orthophthalic acid or maleic acid	[90]
	Fatty acids	Alkylphenol ethoxylates	[81,91]
	hydroxamate (AERO 6493), di-2-(ethylhexyl) phosphoric acid (D2EHPA)	Iso-tridecanol (PX4826)	[76]

3.3. Properties of Flotation Medium

The influence of flotation medium properties on beneficiation efficiency is significant as well, including pH, temperature and ions. For a determined flotation process, the pH and temperature of pulp are usually controlled in a certain range. However, interfering anions and cations are inevitably present, and their concentration varies based on the water quality and solubility of minerals, especially when recycled water is used. To figure out how various anions and cations affect the flotation performance of minerals, extensive efforts have been taken by many researchers. The effect of metal ions on the floatability of apatite was investigated by Ruan et al. [5], and the micro-flotation results were showed in Figure 6. It is seen that the appropriate amount of Ca^{2+} and Mg^{2+} can improve the floatability of apatite but had a negligible effect on the flotation performance of dolomite, whereas Al^{3+}, Fe^{3+} and excessive amounts of Ca^{2+} decreased the recovery of apatite in anionic flotation. The apparent decline of the apatite recovery caused by Al^{3+} and Fe^{3+} was attributed to the preferential precipitation of $Al(OH)_3$ and $Fe(OH)_3$ on the apatite surface, and the hydrophilicity of apatite was enhanced [5].

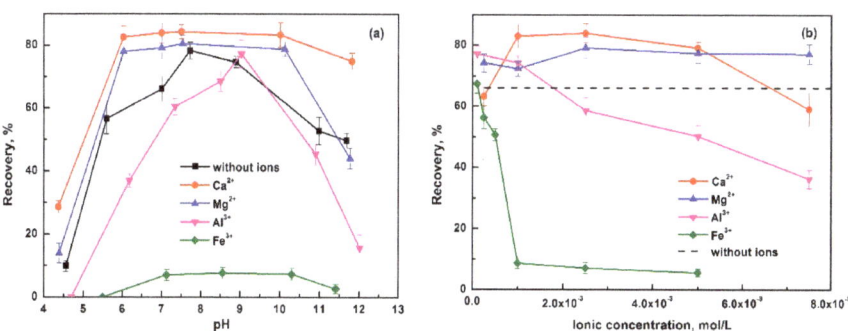

Figure 6. (a) Effect of pH on the floatability of apatite in the presence of metal ions at a concentration of 2.5×10^{-3} M; (b) Effect of ionic concentration on the floatability of apatite at neutral pH [5].

Teague and Lollback found that the recovery of P_2O_5 was inversely proportional to the hardness of the water used in grinding, conditioning and flotation, whilst the P_2O_5 grade of the concentrates was observed to increase from 26.6% to 30.9% as the water hardness decreased [21]. The effects of Ca^{2+}, Mg^{2+}, PO_4^{3-} and SO_4^{2-} on the direct flotation of phosphate ore were investigated by Luo et al. [92]. An adverse influence of these ions on the flotation performance was ascertained, and the selectivity or beneficiation efficiency decreased as ion concentration increased. The authors explained that the Ca^{2+} and Mg^{2+} probably reacted with CO_3^{2-} to form insoluble salts and nonselectively precipitated on the surface of minerals, while PO_4^{3-} and SO_4^{2-} might have interacted with mineral surface through a chemical reaction or adsorption. Consequently, the magnitude of adsorption of the collector decreased, resulting in the efficiency reduction of beneficiation.

The effect of Ca^{2+}, Mg^{2+}, F^- and PO_4^{3-} on the flotation performance of apatite was evaluated by other researchers [16,93]. It was found that all the mentioned ions significantly contributed to the decline in apatite recovery. Santos et al. stated that Ca^{2+} and Mg^{2+} reacted with the apatite collector and caused the reduction of amount available for collection, while F^- reacted with the calcium present in fluorapatite and interfered the collector's function. Hence, it is necessary to minimize the ion concentrations in recycled water by pretreatment due to the detrimental effect on the flotation process caused by various anions and cations [94].

4. Summary

Flotation is dominantly used in the beneficiation of phosphate ores while calcination can be feasibly used in when the fuel energy cost is low and the water source is limited. The application of attrition scrubbing and desliming, magnetic separation and gravity separation should be referred to the characteristic of phosphate ores in chemical composition and texture. In that case, the pre-concentration of valuable minerals and cost reduction are achieved. For example, coarser phosphate concentrates are obtained when use attrition scrubbing, desliming and gravity separation to beneficiate weathered phosphate ores and banded structural sedimentary phosphate ores, respectively. Subsequently, the ultrafine components and overflow products are subjected to flotation for the further recovery of phosphate minerals.

The properties of mineralogy, flotation reagents and the medium have significant impacts on the flotation efficiency or selectivity of phosphate ore. Highly crystallized and coarse-grained phosphate ores have a good processibility. Hence, it is of great necessity to achieve a relatively high degree of mineral liberation and to avoid overgrinding concurrently. Depressants and collectors are the key reagents during the flotation process. Sodium silicate is mostly used to depress silica in a direct flotation. Organic chemical reagents including carboxymethyl cellulose and naphtyl anthyl sulfonates are effective depressants for dolomite. Inorganic acid and phosphoric salts demonstrate a depression effect on the floatability of phosphate minerals. Collectors used in phosphate flotation can be classified into anionic, cationic, amphoteric and mixed surfactants in terms of reagent type. Fatty acids are commonly used as anionic collectors, while cationic amines are used to remove silica in the flotation of phosphate ores. Mixed collectors like mixtures of ionic and nonionic surfactant exhibit more desirable performance than individual collector due to the beneficial effect caused by synergistic advantages. Additionally, the presence of Ca^{2+}, Mg^{2+}, Al^{3+}, Fe^{3+}, F^-, PO_4^{3-} and SO_4^{2-} showed a detrimental effect on the flotation performance of phosphate ores. Consequently, pretreatment should be taken to minimize the ion concentration if there is an excessive amount of interfering anions and cations.

Author Contributions: Writing—original draft preparation, Y.R.; writing—review and revise, D.H. and R.C.

Funding: This research received no external funding.

Conflicts of Interest: The authors declare no conflict of interest.

References

1. U.S. Geological Survey. *Mineral Commodity Summaries*; U.S. Geological Survey: Reston, VA, USA, 2018; pp. 122–123.

2. Abouzeid, A.-Z.M. Physical and thermal treatment of phosphate ores—An overview. *Int. J. Miner. Process.* **2008**, *85*, 59–84. [CrossRef]
3. Sis, H.; Chander, S. Reagents used in the flotation of phosphate ores: A critical review. *Miner. Eng.* **2003**, *16*, 577–585. [CrossRef]
4. Zafar, Z.I.; Anwar, M.M.; Pritchard, D.W. Innovations in beneficiation technology for low grade phosphate rocks. *Nutr. Cycl. Agroecosys.* **1996**, *46*, 135–151. [CrossRef]
5. Ruan, Y.; Zhang, Z.; Luo, H.; Xiao, C.; Zhou, F.; Chi, R. Effects of metal ions on the flotation of apatite, dolomite and quartz. *Minerals* **2018**, *8*, 141. [CrossRef]
6. Zhang, P.; Yu, Y.; Bogan, M. Challenging the "Crago" double float process II. Amine-fatty acid flotation of siliceous phosphates. *Miner. Eng.* **1997**, *10*, 983–994. [CrossRef]
7. Abouzeid, A.-Z.M.; Negm, A.T.; Elgillani, D.A. Upgrading of calcareous phosphate ores by flotation: Effect of ore characteristics. *Int. J. Miner. Process.* **2009**, *90*, 81–89. [CrossRef]
8. Miller, J.D.; Li, M.; Wang, X. Selective Flotation of Phosphate Minerals with Hydroxamate Collectors. U.S. Patent 6341697, 29 January 2002.
9. Ruan, Y.; Zhang, Z.; Luo, H.; Xiao, C.; Zhou, F.; Chi, R. Ambient temperature flotation of sedimentary phosphate ore using cottonseed oil as a collector. *Minerals* **2017**, *7*, 65. [CrossRef]
10. Al-Fariss, T.F.; Ozbelge, H.O.; Abdel Aleem, F.A.; Abdulrazik, S.M. Evaluation of Sandi phosphate rocks for wet process phosphoric acid production. *J. King Saud Univ. Eng. Sci.* **1992**, *4*, 33–44.
11. Li, H.M. Discussion of scrubbing and desliming of phosphate ore in Dianchi area. *Ind. Miner. Process.* **1986**, *6*, 52–53. (In Chinese)
12. He, H.; Yang, W.; Li, R.; Yu, L. Study on flotation of scrubbed tailings from phosphate ore. *Ind. Miner. Process.* **2015**, *11*, 4–10. (In Chinese)
13. Gallala, W.; Herchi, F.; BenAli, I.; Abbassi, L.; Gaied, M.E.; Montacer, M. Beneficiation of phosphate solid coarse waste from Redayef (Gafsa Mining Basin) by grinding and flotation techniques. *Procedia Eng.* **2016**, *138*, 85–94. [CrossRef]
14. Sobhy, A.; Tao, D. Innovative RTS technology for dry beneficiation of phosphate. *Procedia Eng.* **2014**, *83*, 111–121. [CrossRef]
15. Li, N.; Zhang, S.; Peng, H.; Fu, L.; Lu, Y. Application practice and evaluation of photoelectric separation in a phosphate mining industry. *Non Met. Mines* **2018**, *41*, 73–75. (In Chinese) [CrossRef]
16. Guimarães, R.C.; Peres, A.E.C. Interfering ions in the flotation of a phosphate ore in a batch column. *Miner. Eng.* **1999**, *12*, 757–768. [CrossRef]
17. Blazy, P.; Jdid, E.A. Removal of ferriferous dolomite by magnetic separation from the Egyptian Abu Tartur phosphate ore. *Int. J. Min. Process.* **1997**, *49*, 49–58. [CrossRef]
18. Bezzi, N.; Aïfa, T.; Merabet, D.; Pivan, J.-Y. Magnetic properties of the Bled El Hadba phosphate-bearing formation (Djebel Onk, Algeria): Consequences on the enrichment of the phosphate ore deposit. *J. Afr. Earth Sci.* **2006**, *50*, 255–267. [CrossRef]
19. Shaikh, A.M.H.; Dixit, S.G. Beneficiation of phosphate ores using high gradient magnetic separation. *Int. J. Min. Process.* **1993**, *37*, 149–162. [CrossRef]
20. Wei, X.; Huang, Q.; Li, Y. Heavy-media separation industrial production practice of Yichang Huaguoshu Phosphorite. *J. Wuhan Inst. Tech.* **2011**, *33*, 48–52. (In Chinese)
21. Teague, A.J.; Lollback, M.C. The beneficiation of ultrafine phosphate. *Min. Eng.* **2012**, *27–28*, 52–59. [CrossRef]
22. Zafar, Z.I.; Anwar, M.M.; Pritchard, D.W. Optimization of thermal beneficiation of a low grade dolomitic phosphate rock. *Int. J. Miner. Process.* **1995**, *43*, 123–131. [CrossRef]
23. Watti, A.; Alnjjar, M.; Hammal, A. Improving the specifications of Syrian raw phosphate by thermal treatment. *Arab. J. Chem.* **2016**, *9*, 637–642. [CrossRef]
24. Özer, A.K. The characteristics of phosphate rock for upgrading in a fluidized bed. *Adv. Powder Technol.* **2003**, *14*, 33–42. [CrossRef]
25. Zafar, Z.I.; Ashraf, M. Selective leaching kinetics of calcareous phosphate rock in lactic acid. *Chem. Eng. J.* **2007**, *131*, 41–48. [CrossRef]
26. Zafar, Z.I.; Anwar, M.M.; Pritchard, D.W. Selective leaching of calcareous phosphate rock in formic acid: Optimisation of operating conditions. *Min. Eng.* **2006**, *19*, 1459–1461. [CrossRef]

27. Ashraf, M.; Zafar, I.Z.; Ansari, T.M. Selective leaching kinetics and upgrading of lowgrade calcareous phosphate rock in succinic acid. *Hydrometallurgy* **2005**, *80*, 286–292. [CrossRef]
28. Abu-Eishah, S.I.; Muthaker, M.; Touqan, N. A new technique for the beneficiation of low grade carbonate-rich phosphate rocks by digestion with dilute acetic acid solutions: Pilot plant testing results. *Miner. Eng.* **1991**, *4*, 573–586. [CrossRef]
29. Gharabaghi, M.; Noaparast, M.; Irannajad, M. Selective leaching kinetics of low-grade calcareous phosphate ore in acetic acid. *Hydrometallurgy* **2009**, *95*, 341–345. [CrossRef]
30. Gharabaghi, M.; Irannajad, M.; Noaparast, M. A review of the beneficiation of calcareous phosphate ores using organic acid leaching. *Hydrometallurgy* **2010**, *103*, 96–107. [CrossRef]
31. Rodrigues, A.J.; Brandao, P.R.G. The influence of crystal chemistry properties on the floatability of apatites with sodium oleate. *Miner. Eng.* **1993**, *6*, 643–653. [CrossRef]
32. Santana, R.C.; Duarte, C.R.; Ataide, C.H.; Barrozo, M.A.S. Flotation selectivity of phosphate ore: Effect of particle size and reagent concentration. *Sep. Sci. Technol.* **2011**, *46*, 1511–1518. [CrossRef]
33. Santana, R.C.; Farnese, A.C.C.; Fortes, M.C.B.; Ataide, C.H.; Barrozo, M.A.S. Influence of particle size and reagent dosage on the performance of apatite flotation. *Sep. Sci. Technol.* **2008**, *64*, 8–15. [CrossRef]
34. Guo, F.; Li, J. Separation strategies for Jordanian phosphate rock with siliceous and calcareous gangues. *Int. J. Miner. Process.* **2010**, *97*, 74–78. [CrossRef]
35. Al-Wakeela, M.I.; Lin, C.L.; Miller, J.D. Significance of liberation characteristics in the fatty acid flotation of Florida phosphate rock. *Miner. Eng.* **2009**, *22*, 244–253. [CrossRef]
36. Qun, W.; Heiskanen, K. Batch flotation tests by fatty acid on a phosphate-iron oxide-silicate regolith ore sample from Sokli, Finland. *Miner. Eng.* **1990**, *3*, 473–481. [CrossRef]
37. Silva, J.P.P.; Baltar, C.A.M.; Gonzaga, R.S.G.; Peres, A.E.C.; Leite, J.Y.P. Identification of sodium silicate species used as flotation depressants. *Min. Met. Process.* **2012**, *29*, 207–210. [CrossRef]
38. Rao, D.S.; Vijayakumar, T.V.; Angadi, S.; Prabhakar, S.; Raju, G.B. Effects of modulus and dosage of sodium silicate on limestone flotation. *Maejo Int. J. Sci. Technol.* **2010**, *4*, 397–404.
39. Sayilgan, A.; Arol, A.I. Effect of carbonate alkalinity on flotation behavior of quartz. *Int. J. Miner. Process.* **2004**, *74*, 233–238. [CrossRef]
40. Pavlovic, S.; Brandao, P.R.G. Adsorption of starch, amylose, amylopectin and glucose monomer and their effect on the flotation of hematite and quartz. *Min. Eng.* **2003**, *16*, 1117–1122. [CrossRef]
41. Guimarães, R.C.; Araujo, A.C.; Peres, A.E.C. Reagents in igneous phosphate ores flotation. *Min. Eng.* **2005**, *18*, 199–204. [CrossRef]
42. Zheng, X.; Smith, R.W. Dolomite depressants in the flotation of apatite and collophane from dolomite. *Min. Eng.* **1997**, *10*, 537–545. [CrossRef]
43. Yu, J.; Ge, Y.; Guo, X.; Guo, W. The depression effect and mechanism of NSFC on dolomite in the flotation of phosphate ore. *Sep. Purif. Technol.* **2016**, *161*, 88–95. [CrossRef]
44. Zheng, X.; Arps, P.J.; Smith, R.W. Adhesion of two bacteria onto dolomite and apatite: Their effect on dolomite depression in anionic flotation. *Int. J. Min. Process.* **2001**, *62*, 159–172. [CrossRef]
45. Somasundaran, P.; Wang, D.Z. *Solution Chemistry: Minerals and Reagents*, 1st ed.; Elsevier: Amsterdam, The Netherlands, 2006; pp. 58–60.
46. Elgillani, D.A.; Abouzeid, A.-Z.M. Flotation of carbonates from phosphate ores in acidic media. *Int. J. Min. Process.* **1993**, *38*, 235–256. [CrossRef]
47. Chen, Y.; Feng, Q.; Zhang, G.; Liu, D.; Liu, R. Effect of sodium pyrophosphate on the reverse flotation of dolomite from apatite. *Minerals* **2018**, *8*, 278. [CrossRef]
48. Al-Fariss, T.F.; Arafat, Y.; Abd El-Aleem, F.A.; El-Midany, A.A. Investigating sodium sulphate as a phosphate depressant in acidic media. *Sep. Purif. Technol.* **2014**, *124*, 163–169. [CrossRef]
49. El-Mofty, S.E.; El-Midany, A.A. Role of calcium ions and their interaction with depressants in phosphate flotation. *Chem. Pap.* **2018**, *72*, 2641–2646. [CrossRef]
50. Zhang, P.; Snow, R. Evaluation of phosphate depressants in the phosphate/silica system. *Miner. Metall. Process.* **2009**, *26*, 101–104. [CrossRef]
51. Nagaraj, D.R.; Rothenberg, A.S.; Lipp, D.W.; Panzer, H.P. Low molecular weight polyacrylamide-based polymers as modifiers in phosphate beneficiation. *Int. J. Min. Process.* **1987**, *20*, 291–308. [CrossRef]
52. Dho, H.; Iwasaki, I. Role of sodium silicate in phosphate flotation. *Min. Metallurgy Explor.* **1990**, *7*, 215.

53. Clifford, P.R.; Lloyd, G.M.; Zhang, J.P.; Richardson, S.G.; Birky, B.K.; Stewart, K.J. *An Investigation of Flotation Reagents*; Florida Institute of Phosphate Research: Bartow, FL, USA; p. 2008.
54. Nagaraj, D.R.; Rothenberg, A.S.; Lambert, A.S. Flotation Beneficiation Process for Non-Sulfide Minerals. U.S. Patent 4720339, 19 January 1988.
55. Ribeiro, R.C.C.; Correia, J.C.G.; Monte, M.B.M.; Seidl, P.R.; Mothe, C.G.; Lima, C.A. Cashew gum: A new depressor for limestone in the phosphate minerals flotation. *Miner. Eng.* **2003**, *16*, 873–875. [CrossRef]
56. Hanna, J.; Anazia, I. Fatty acid separation of siliceous carbonate phosphates. *Miner. Metall. Process.* **1990**, *10*, 84–90. [CrossRef]
57. Lehr, J.R.; Hsieh, S. Beneficiation of High Carbonate Phosphate Ores. U.S. Patent 4287053, 1 September 1981.
58. Hsieh, S.; Lehr, J.R. Method of Beneficiating High Carbonate Phosphate Ore. U.S. Patent 4486301, 4 December 1984.
59. Rao, D.V.; Narayanan, M.K.; Nayak, U.B.; Ananthapadmanabhan, K.; Somasundaran, P. Flotation of calcareous Mussorie phosphate ore. *Int. J. Miner. Process.* **1985**, *14*, 57–66. [CrossRef]
60. Yehia, A.; Khalek, M.A.; Ammar, M. Cellulase as a new phosphate depressant in dolomite-phosphate flotation. *Physicochem. Probl. Miner. Process.* **2017**, *53*, 1092–1104.
61. Smani, S.M. Process for Enrichment by Flotation of Phosphate Ores with Gangues Containing Carbonates. U.S. Patent 4008151, 15 February 1977.
62. Mohammadkhani, M.; Noaparast, M.; Shafaei, S.Z.; Amini, A.; Amini, E.; Abdollahi, H. Double reverse flotation of a very low grade sedimentary phosphate rock, rich in carbonate and silicate. *Int. J. Miner. Process.* **2011**, *100*, 157–165. [CrossRef]
63. Paiva, P.R.P.; Monte, M.B.M.; Simao, R.A.; Gaspar, J.C. In situ AFM study of potassium oleate adsorption and calcium precipitate formation on an apatite surface. *Miner. Eng.* **2011**, *24*, 387–395. [CrossRef]
64. Santos, E.P.; Dutra, A.J.B.; Oliveira, J.F. The effect of jojoba oil on the surface properties of calcite and apatite aiming at their selective flotation. *Int. J. Miner. Process.* **2015**, *143*, 34–38. [CrossRef]
65. Huang, Q.; Huang, J.; Zhou, H.; Pan, Z.; Ping, X. Synthesis and application of a flotation collector for collophanite. In *Beneficiation of Phosphates: New Thought, New Technology, New Development*; Zhang, P., Miller, J., El-Shall, H., Eds.; SME: Denver, CO, USA, 2012; pp. 359–364.
66. Oliveira, M.S.; Santana, R.C.; Ataíde, C.H.; Barrozo, M.A.S. Recovery of apatite from flotation tailings. *Sep. Purif. Technol.* **2011**, *79*, 79–84. [CrossRef]
67. Guo, F.; Li, J. Selective separation of silica from a siliceous-calcareous phosphate rock. *Min. Sci. Technol. (China)* **2011**, *21*, 135–139.
68. Sahoo, H.; Rath, S.S.; Das, B.; Mishra, B.K. Flotation of quartz using ionic liquid collectors with different functional groups and varying chain lengths. *Miner. Eng.* **2016**, *95*, 107–112. [CrossRef]
69. Vieira, A.M.; Peres, A.E.C. The effect of amine type, pH, and size range in the flotation of quartz. *Min. Eng.* **2007**, *20*, 1008–1013. [CrossRef]
70. Nunes, A.P.L.; Pinto, C.L.L.; Valadao, G.E.S.; Viana, P.R.M. Floatability studies of wavellite and preliminary results on phosphorus removal from a Brazilian iron ore by froth flotation. *Miner. Eng.* **2012**, *39*, 206–212. [CrossRef]
71. Shao, X.; Jiang, C.L.; Parekh, B.K. Enhanced flotation separation of phosphate and dolomite using a new amphoteric collector. *Miner. Metall. Process.* **1998**, *15*, 11–14. [CrossRef]
72. Abdel-Khalek, M. Separation of dolomite from phosphate minerals by flotation with a new amphoteric surfactant as collector. *Trans. Inst. Min. Met. C* **2001**, *110*, 89–93.
73. Elmahdy, A.M.; El-Midany, A.A.; Abdel-Khalek, N.A. Application of amphoteric collector for dolomite separation by statistically designed experiments. *Trans. Inst. Min. Met. C* **2007**, *116*, 72–76. [CrossRef]
74. Hu, Y.; Wang, D. Flotation of apatite and calcite using α-amino aryl phosphoric acid as collector. *Nonferrous Met.* **1992**, *44*, 41–43. (In Chinese)
75. Hu, Y.; Xu, Z. Interactions of amphoteric amino phosphoric acids with calcium-containing minerals and selective flotation. *Int. J. Miner. Process.* **2003**, *72*, 87–94. [CrossRef]
76. Filippova, I.V.; Filippov, L.O.; Duverger, A.; Severov, V.V. Synergetic effect of a mixture of anionic and nonionic reagents: Ca mineral contrast separation by flotation at neutral pH. *Min. Eng.* **2014**, *66–68*, 135–144. [CrossRef]
77. Rao, K.H.; Forssberg, K.S.E. Mixed collector systems in flotation. *Int. J. Miner. Process.* **1997**, *51*, 67–79.

78. Javadian, S.; Gharibi, H.; Bromand, Z.; Sohrabi, B. Electrolyte effect on mixed micelle and interfacial properties of binary mixtures of cationic and nonionic surfactants. *J. Colloid Interface Sci.* **2008**, *318*, 449–456. [CrossRef]
79. Szymczyk, K.; Janczuk, B. The adsorption at solution-air interface and volumetric properties of mixtures of cationic and nonionic surfactants. *Colloids Surf. A Phys. Eng. Asp.* **2007**, *293*, 39–50. [CrossRef]
80. Sis, H.; Chander, S. Adsorption and contact angle of single and binary mixtures of surfactants on apatite. *Min. Eng.* **2003**, *16*, 839–848. [CrossRef]
81. Sis, H.; Chander, S. Improving froth characteristics and flotation recovery of phosphate ores with nonionic surfactants. *Miner. Eng.* **2003**, *16*, 587–595. [CrossRef]
82. Wang, X.; Nguyen, A.V.; Miller, J.D. Selective attachment and spreading of hydroxamic acid–alcohol collector mixtures in phosphate flotation. *Int. J. Miner. Process.* **2006**, *78*, 122–130. [CrossRef]
83. Filippov, L.O.; Duverger, A.; Filippova, I.V.; Kasaini, H.; Thiry, J. Selective flotation of silicates and Ca-bearing minerals: The role of non-ionic reagent on cationic flotation. *Min. Eng.* **2012**, *36–38*, 314–323. [CrossRef]
84. Rao, K.H.; Dwari, R.K.; Lu, S.; Vilinska, A.; Somasundaran, P. Mixed anionic/non-ionic collector in phosphate gangue flotation from magnetite fines. *Open Miner. Process. J.* **2011**, *4*, 14–24. [CrossRef]
85. Hughes, C.V. Enhanced Flotation Reagents for Beneficiation of Phosphate Ores. U.S. Patent 5962828, 5 October 1999.
86. Preller, G.S.; Schoeman, B.J.K. Flotation of Apatite. U.S. Patent 3405802, 15 October 1968.
87. Cao, Y.; Huang, G.; Yang, L.; Liu, S.; Deng, Q.; Gu, X. Process study on some low grade collophanite with high content of magnesium. *Non Met. Mines* **2016**, *39*, 63–66. (In Chinese)
88. Hefner, R.E., Jr. N-Aminoethylpiperazine Condensates for Beneficiation of Phosphate Ore. U.S. Patent 4301004, 17 November 1981.
89. Subrahmanyam, C. Method for Conditioning Phosphate Ores. U.S. Patent 4556545, 12 March 1985.
90. Dorrepaal, W.; Haak, G.M. Process for the Flotation of Ores. U.S. Patent 4200522, 29 April 1980.
91. Giesekke, E.W.; Harris, P.J. The role of polyoxyethylene alkyl ethers in apatite flotation at foskor, phalaborwa (South Africa). *Min. Eng.* **1994**, *7*, 1345–1361. [CrossRef]
92. Luo, H.; Liu, L.; Xie, B.; Ou, S. Effect of Ca^{2+}, Mg^{2+}, PO_4^{3-} and SO_4^{2-} on the flotation of phosphate. In *Beneficiation of Phosphates: New Thought, New Technology, New Development*; Zhang, P., Miller, J., El-Shall, H., Eds.; SME: Denver, CO, USA, 2012; pp. 33–38.
93. Santos, M.A.; Santana, R.C.; Capponi, F.; Ataíde, C.H.; Barrozo, M.A.S. Effect of ionic species on the performance of apatite flotation. *Sep. Purif. Technol.* **2010**, *76*, 15–20. [CrossRef]
94. Nanthakumar, B.; Grimm, D.; Pawlik, M. Anionic flotation of high-iron phosphate ores—Control of process water chemistry and depression of iron minerals by starch and guar gum. *Int. J. Miner. Process.* **2009**, *92*, 49–57. [CrossRef]

© 2019 by the authors. Licensee MDPI, Basel, Switzerland. This article is an open access article distributed under the terms and conditions of the Creative Commons Attribution (CC BY) license (http://creativecommons.org/licenses/by/4.0/).

Article

The Beneficiation Processes of Low-Grade Sedimentary Phosphates of Tozeur-Nefta Deposit (Gafsa-Metlaoui Basin: South of Tunisia)

Haïfa Boujlel [1,2,*], Ghassen Daldoul [1,2], Haïfa Tlil [1,2], Radhia Souissi [2], Noureddine Chebbi [3], Nabil Fattah [4] and Fouad Souissi [1,2]

1. Department of Geology, Faculty of Sciences of Tunis, University of Tunis El Manar, Tunis 2092, Tunisia; daldoul.ghassen@yahoo.com (G.D.); haifatlili1@hotmail.fr (H.T.); souissifoued2@gmail.com (F.S.)
2. National Institute of Research and Physico-Chemical Analysis (INRAP), Technopole of Sidi Thabet, Ariana 2020, Tunisia; souissiradhia@yahoo.fr
3. National Office of Mines (ONM), Tunis 2035, Tunisia; noureddine.chebbi23@yahoo.com
4. Research Centre of the Phosphate Company of Gafsa (CPG), Metlaoui 2130, Tunisia; nabil.fattah@cpg.com.tn
* Correspondence: boujlelhaifa@yahoo.fr; Tel: +216-95-765-104

Received: 22 October 2018; Accepted: 17 December 2018; Published: 20 December 2018

Abstract: The enrichment of the low-grade-phosphate ore of the Tozeur-Nefta deposit was investigated using scrubbing-attrition, ball grinding and anionic/cationic reverse flotation in order to separate phosphate-rich particles from their gangue. The choice of the beneficiation process was based on the petrographic, mineralogical and chemical analyses. The petrographic and mineralogical studies have revealed the abundance of phosphatic (carbonate-fluorapatite-CFA) particles) coupled with carbonates (calcite, dolomite,) and silicates (quartz, illite, kaolonite) that constitute the (endo/exo) gangue of these ores. Chemical analysis has demonstrated that the raw phosphate sample contains low amounts of P_2O_5 (12.0%) and MgO (4.9%) and high amounts of CaO (40.7%) and SiO_2 (20.5%). Microscopic observation/counting has shown that the release mesh occurs in the 71–315 μm size. Scrubbing-attrition, grinding and reverse flotation methods were applied to the +71μm fraction. Scrubbing-attrition tests of the 71–315 μm fraction have helped to improve the P_2O_5 grade to 15.5%. Ball-grinding tests were used to reduce the coarse fraction +315 μm. Grounded materials were sieved to 71–315 μm and combined with the scrubbed fraction in the flotation feed. Reverse-flotation tests of the phosphate-rich fraction (71–315 μm) have helped to improve the P_2O_5 grade to 27.1%, with a recovery rate of 92.4%.

Keywords: Gafsa-Metlaoui Basin; low-grade phosphate; silicate-carbonate gangue; froth flotation; upgrading

1. Introduction

Tunisia is the second largest phosphate producer in Africa, with an annual production of about 8 million tons (Mt) [1,2]. The country also boasts about 1750 Mt of phosphate reserves. The main phosphate resources in Tunisia are distributed over three major Eocene sedimentary basins: the Gafsa-Metlaoui Basin, the Meknessi Basin and the Sra Ouertane Basin [3–6]. The Tozeur-Nefta deposit, which is the subject of the current study, is located in the Southwest region of the Gafsa-Metlaoui Basin, and has remained untapped. In order to upgrade its phosphate-production capacity, the Compagnie des Phosphates de Gafsa (CPG, Tunisia) has planned to make use of the Tozeur-Nefta deposit before 2020. This deposit, which has been the subject of several studies [7–11], contains about 410 Mt of phosphate, has been considered, to date, as a strategic reserve. The majority of the low-grade phosphate ore is organized either in the form of a single layer or a layer-beam including non-phosphatic intercalations [12].

Currently, phosphate fertilizers are generally produced from primary-phosphate ores. Phosphate material can easily be separated from the gangue and concentrated by conventional beneficiation techniques (i.e. washing process). As a result of the concurrent increase in demand for phosphate products, and the continuous depletion of phosphate reserves, much more focus is made today on opportunities to recover phosphate ores from secondary sources (i.e. tailings and low-grade ore deposits). These new sources are continually investigated, as is happening now at the Hazara deposit in Pakistan (16% P_2O_5) and the Kohe-lar deposit in Iran (5.01% P_2O_5) [13,14]. Upgrading processes of sedimentary-type phosphate ores consist of separating the gangue minerals (silicates and carbonates) from the valuable phosphate material [15–17]. New processing techniques are needed to upgrade the low-grade phosphates to commercial-grade products (P_2O_5 > 28%) [18]. The choice of the suitable beneficiation process requires a good knowledge of the mineral assemblages of the studied ore [19]. Several industrial processes are used worldwide for the enrichment of phosphate ores, ranging from simple mechanical preparation to more complex treatment schemes, which combine several methods such as flotation, magnetic, electrostatic and dense-media separation (DMS). Calcination is applied for the enrichment of phosphates that are rich in organic materials, like those of the Youssoufia deposit in Morocco [12], or that include a carbonated gangue, like those of Akashart's phosphates in Iraq [20]. Electromagnetic separation is used mainly for the valorization of igneous phosphates, such as that of the Phalaborawa deposit in South Africa [21]. Acid leaching is a process of enrichment of siliceous and calcareous phosphate ore applied to the Abbotabad deposit in Pakistan [22,23]. Magnetic separation is used to separate magnetic minerals from phosphate ore of the feebly magnetic ankerite, as noted in the Abu Tartur phosphate ore deposit of Egypt [24]. Practically speaking, crushing, grinding and screening methods are used to remove coarse-textured materials. These techniques are also used in the large phosphate-ore deposits of Kef Eddour and Oum El Khecheb, located in the Tunisian Gafsa-Metlaoui Basin [25,26]. Generally, washing, attrition and desliming methods are used to remove the major part of the clay-type binder [27].

Flotation is the most widely used method for the enrichment of medium-grade phosphate rock. This method is mainly employed for the separation of phosphate minerals from silicate gangue [28,29]. Flotation may be conducted in of two ways depending on the composition of the material to be processed: direct flotation and/or reverse flotation. That is to say, direct flotation consists of floating the phosphate minerals and depressing the gangue material, while reverse flotation consists of floating the gangue minerals and keeping the phosphatic matter in the slurry. Direct flotation is less effective in the presence of carbonated gangue because of similar physicochemical properties between phosphate and carbonate minerals [28]. Reverse flotation is the most common method used for the separation of phosphate minerals from carbonated gangue [30].

Several studies on reverse-flotation of phosphates with siliceous and carbonated gangues showed satisfactory results in terms of grade and recovery of P_2O_5 [14,26,31–36]. Reverse flotation can be applied as unique process for the beneficiation of sedimentary siliceous carbonate phosphates, such as those of Ayata (Tunisia) [31] and Sebaiya (Egypt) [33]. It can be preceded by other techniques including grinding (e.g. igneous phosphates of kola (Brazil) [32] and sedimentary phosphates of Kef-Eddur (Tunisia) [26]) and gravity separation followed by direct flotation (e.g. phosphate ore slime of Yichang (China) [36]). Other flotation parameters have been studied; the most important are release mesh, flotation reagents type/dosage, and pH and pulp concentration. The present work differs from previous studies by in the following way: (i) release mesh, (ii) processes that precede flotation (sizing, washing, attrition and ball milling), (iii) two reverse flotation stages, although used by others, it was performed in this work under specific conditions of pH, reagents type/dosage and air flow-rate.

The present study consists of characterizing the low-grade sedimentary phosphate ore of the roof series of the Tozeur-Nefta deposit, while having for objectives: (i) the examination of the mineralogical and geochemical characteristics of the studied phosphate ore, and (ii) the design of an efficient and economic process for the beneficiation of low-grade phosphate ores, including siliceous and carbonated gangues.

2. Materials and Methods

2.1. Sampling and Separation

The trench sampling method was used to extract 400 kg of raw phosphate material from the roof series of the Tozeur-Nefta deposit. All samples were crushed using a jaw crusher to 100% below 10 mm in size, then homogenized and divided by a riffle-sample divider to obtain representative sub-samples, each one weighing 2 kg. After further quartering, one sub-sample was ground manually in an agate mortar into a fine powder for XRD and chemical analysis. One of the non-ground sub-samples was used for grain-size analysis as well as for microscopic observation. The remaining sub-samples were stored in plastic bags for phosphate beneficiation process.

2.2. Analysis

A representative sub-sample from the roof phosphate layer of the Tozeur-Nefta deposit was used in this study to conduct the mineralogical and geochemical grain-size analyses.

2.2.1. Petrographic and Mineralogical Analysis

The petrographic analysis was carried out by means of thin sections, which were examined and photographed with an optical microscope. Mineralogical characterization was carried out on the basis of an XRD analysis and binocular observation methods. XRD analyses were carried out by means of PANalytical X'Pert PRO X-ray diffractometer (Cu Kα radiation; λ = 1.5418 Å, PANalytical, Almelo, The Netherland). Mineralogical identification was performed by means of High Score Plus software (version 2.1, PANalytical, Almelo, The Netherland) equipped with the ICDD PDF–2 Release 2004.

2.2.2. Particle-Size Analysis and Heavy-Liquid Separation

Particle-size distribution was measured by means of the wet-sieve (AFNOR) method (from 0.40 to 10 mm size range). Binocular observation of the different size fractions of the sample was used to identify and quantify, by visual inspection and grain counting, the relative percentages of the major minerals, and to evaluate the release mesh of phosphate minerals. Dense media separation was evaluated at laboratory scale through heavy-liquid separation (HLS) test work. Tetra Bromo Ethane (TBE, density (d) = 2.96 g/cm^3) was used to separate phosphate minerals (d > 2.96 g/cm^3) from silicate and carbonated gangue minerals (d < 2.96 g/cm^3). Phosphate fraction was dried and examined by binocular observation in order to manually sort phosphate grains from the remaining impurities.

2.2.3. Chemical Analysis

Chemical and analytical procedures, applied to the studied phosphate material, as considered in the present study, are adapted from the protocols of CPG and tested in the company's labs, where standardized methods are used specifically for total digestion and chemical analysis of sedimentary phosphate rocks. P_2O_5 and CaO were analyzed by colorimetry (Technicon Auto-analyzer, Bran-Luebbe, SPX Process Equipment, Norderstedt, Germany) after digestion by $HClO_4$-HCl. Atomic adsorption spectroscopy (Perkin Elmer-AAnalyst 800, Perkin Elmer, Norwalk, CT, USA) was used to analyze MgO, Fe_2O_3, Al_2O_3, Cu, Zn, Mn and Ni after digestion by $HClO_4$-HF-HNO_3, SiO_2 after digestion by HCl-HF and Cd after digestion by HNO_3. K_2O and Na_2O were analyzed by flame spectrometry after digestion by $HClO_4$-HF-HNO_3. SO_3 was analyzed by gravimetry after digestion by HCl and subsequent precipitation of SO_4^{2-} ions as $BaSO_4$, with a $BaCl_2$. Loss-on-ignition (LOI) was determined by means of a muffle furnace at 1050 °C. To verify accuracy, standard reference phosphate material SRM694 (Western phosphate rock) and SRM120C (Florida phosphate rock) were digested and analyzed according to the same protocol applied to the phosphate material considered in this study. All chemical analyses presented experimental errors of less than 5%.

2.3. Beneficiation Processes

Beneficiation will be conducted to upgrade the +71 μm size fraction by means of sizing, scrubbing-attrition, grinding and flotation methods.

2.3.1. Sizing, Scrubbing (Washing) and Attrition

Scrubbing treatment of raw phosphate (2 kg) was carried out for 15 min (70 rpm) and by using a 10-L scrubber and with 40% solids in order to remove clay impurities from the surface of grains. After scrubbing, the resulting pulp was wet-sieved (AFNOR sieves, see Section 2.2.2) to obtain the 71–315 μm fraction and then subjected to attrition in order to remove further clay coatings from phosphate grains. An attrition step was conducted in a Denver cell at 60% solid/liquid ratio, respectively, and 1500 rpm (5, 10 and 15 min). The product from attrition was wet-sieved to eliminate the fraction −71 μm and the 71–315 μm fraction was used as a flotation feed.

2.3.2. Grinding

In general, the grinding of ores may lead to the liberation of the valuable minerals from the gangue. It is obvious that wet grinding often requires less energy than dry grinding [37]. The wet-grinding process was investigated for the +315 μm fraction in order to improve the liberation of phosphate grains from the gangue and to make this coarse fraction more suitable for flotation. The characteristics of grinding charge are displayed in Table 1. Steel balls of four types (B1, B2, B3 and B4), of different diameters and weights were used in the study. The large balls are suitable for the fractionation of coarse grains and small balls for small grains [38]. A grinding process was carried out in a ball mill of 10-L capacity. The influence of three grinding parameters on the product fineness was studied and optimized: (i) solid concentration (30%, 40% and 50%), (ii) grinding time (3, 4 and 5 min) and (iii) charge ratio (3/1, 4/1 and 5/1) (Table 1). The rotation speed was fixed at 50 rpm.

Table 1. Characteristics of grinding balls.

Grinding Ball	Diameter (mm)	Ball Ratio (%)	Weight (kg)
B1	40	7.06	0.29
B2	30	55.5	0.12
B3	25	8.91	0.07
B4	20	28.5	0.04

2.3.3. Flotation

Reverse flotation was carried out by means of two flotation reagents kindly supplied by Clariant® Company (Shanghai, China): (i) V2711 Flotinor ([R–O–(CH$_2$)$_3$–NH$_3$]+CH$_3$COO–; anionic collector) was used as calcite collector and (ii) Ether diamine Flotigam 2835-2L ([R–O–(CH$_2$)$_3$-NH–(CH$_2$)$_3$–NH$_3$]+CH$_3$COO–; cationic collector) as quartz collector [39–41]. Fresh solutions of these reagents were prepared daily to be used in flotation tests. Dilute H$_3$PO$_4$ (5%) acted both as phosphate depressant and pH-modifier. Reverse flotation tests were performed in a laboratory Denver 1.5 L flotation cell (model D-12). For each flotation test, a 500 g of phosphate sample was used. The dilution of the pulp was fixed at 35% solid, rotation speed at 1500 rpm and air flow-rate at 5.4 L/min. Phosphate is recovered after two stages of reverse flotation: first, separating phosphate from silicate gangue minerals in natural pH (pH = 7.8) by means of cationic and anionic collectors, and then, by means of anionic and H$_3$PO$_4$ reagents in acid medium (pH = 5.0) to separate phosphate from carbonate minerals. Under acid condition, phosphate minerals are not taken up by the foams and remain in the cell.

3. Results and Discussion

3.1. Petrographic Analysis

The most abundant phosphatic elements are displayed as rounded to sub-rounded pellets embedded in microsparitic cement (Figure 1a). These elements are often surrounded by a thin and clear cortex (Figure 1b). Other elements were observed in thin sections, such as bone debris of elongated shapes (Figure 1c) and ooliths formed by growth and crystallization of micrometric phosphate layers (Figure 1d). Aggregates of small dolomitic rhombohedra (Figure 1e) constitute the carbonated bonding phase, which generally indicates the beginning of crystallization of the micritic material containing all the impurities of the original mud. The exogangue is generally carbonated, showing phosphate grains surrounded by dolomite rhombohedra (Figure 1f), while the endogangue is present either as silica (Figure 1g) or carbonate grains (Figure 1h), occurring, respectively, as quartz or calcite included within the phosphate grains.

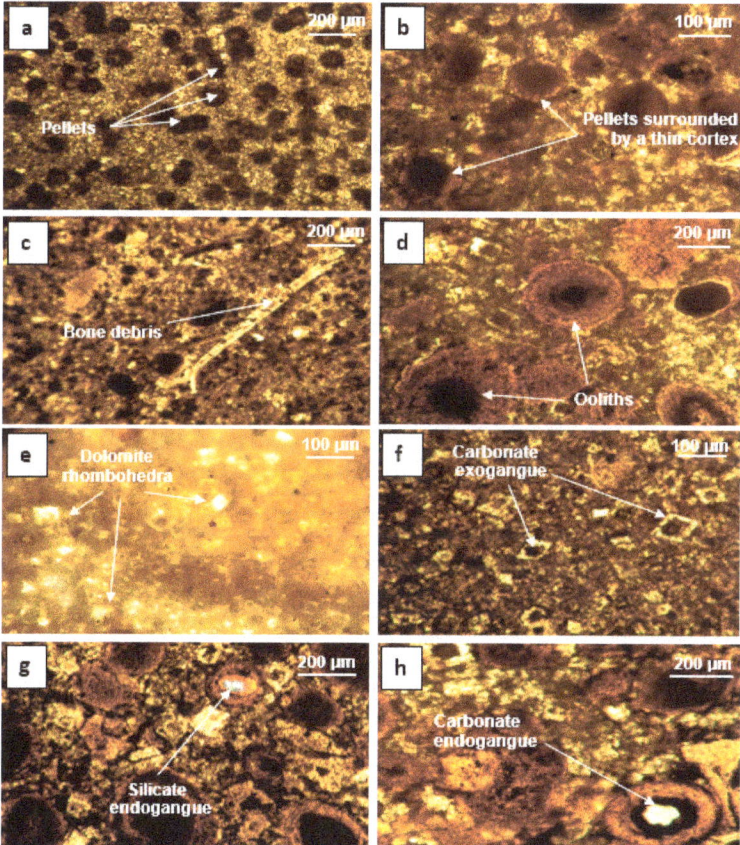

Figure 1. Optical microscope observation (plane polarized light) of the roof phosphates of the Tozeur-Nefta deposit. (**a**) Pellets of rounded to sub-rounded shapes bound by a microsparitic cement; (**b**) Pellets surrounded by a thin clear cortex; (**c**) Bone debris of elongated forms; (**d**) Oolites formed by the deposition of successive micrometric layers of phosphate; (**e**) Carbonate bonding phase formed by euhedral dolomite; (**f**) Carbonate exogangue made of euhedral dolomite; (**g**) Silicate (quartz) endogangue making the nucleus of the phosphate grain; (**h**) Carbonate (calcite) endogangue making the nucleus of the phosphate grain.

3.2. Mineralogical Analysis

The XRD pattern (Figure 2) revealed the presence of seven mineral species, as follows: carbonate-fluorapatite (CFA), calcite, dolomite, illite, kaolinite, gypsum, and quartz. The binocular observation revealed (Figure 3) the abundance of phosphate elements in the 71–315 µm fractions, carbonate minerals are more concentrated in the −200 µm fraction, whereas free silica is mainly concentrated in the −160 µm fraction. The mixed grains are concentrated in the coarse fraction (+315 µm). The release mesh was defined by Blazy [42], as the size that the grains should have for the best detachment of ore particles from the gangue (mixed grain ≤ 10%). The release mesh of phosphate grains, determined by binocular microscope, corresponds to the 71–315 µm fractions (Figure 3).

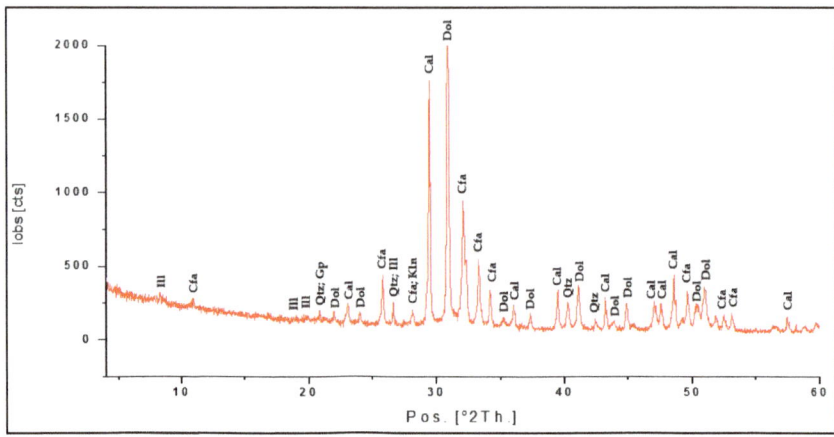

Figure 2. X-ray diffractogram of the raw phosphate sample showing the presence of: carbonate-fluorapatite (Cfa), calcite (Cal), dolomite (Dol), illite (Ill), kaolinite (Kln), gypsum (Gp) and quartz (Qtz).

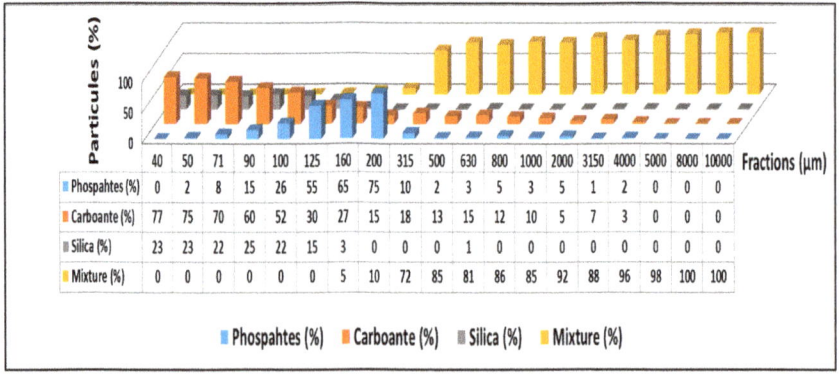

Figure 3. Distribution of the main minerals in different size fractions of the raw phosphate sample.

3.3. Chemical Analysis

3.3.1. Raw Phosphate Analysis

Chemical analysis (Table 2) has shown a low concentration of P_2O_5 (12.0%) in the representative phosphate sample of the roof series of Tozeur-Nefta deposit. It is worth noting that phosphate ores can be classified into three groups depending on their P_2O_5 content [43]: (i) high-grade phosphate ore (26 to 35% P_2O_5), (ii) moderately low-grade phosphate ore (17 to 25% P_2O_5), and (iii) low-grade phosphate ore (12 to 16% P_2O_5). According to this classification, the phosphate ore of the top series of the Tozeur-Nefta deposit belongs to the class of low-grade phosphate. The CaO/P_2O_5 ratio (3.39) is much higher than that of pure CFA (1.55) [4–6]. The concentration of SiO_2 was very high (20.5%). The concentration of Fe_2O_3 and Al_2O_3 amounting 0.65% and 1.75%, respectively, give a ratio ((Fe_2O_3 + Al_2O_3)/P_2O_5) of 0.2. It is worth noting that for values > 0.1, concentrations of Fe_2O_3 and Al_2O_3 are considered as high and, therefore, many problems can arise during the filtration and purification of phosphoric acid [44]. Organic-matter (0.28% organic carbon), Na_2O (0.39%) and K_2O (0.45%) contents are relatively low.

Results of trace elements concentrations are provided in Table 2. The measured Ni (80 mg/kg), Zn (215 mg/kg), Cd (65 mg/kg), Cu (31 mg/kg) and Mn (296 mg/kg) concentrations are in agreement with their equivalents in the unwashed (raw) economic ores from the main phosphate series of the Gafsa-Metlaoui basin [1,4,5,45,46].

Table 2. Concentrations of major and trace elements in the studied raw phosphate.

Element	Concentration
P_2O_5 (%)	12.0
CaO (%)	40.7
MgO (%)	4.95
SiO_2 (%)	20.5
SO_3 (%)	1.27
Fe_2O_3 (%)	0.65
Al_2O_3 (%)	1.75
Na_2O (%)	0.39
K_2O (%)	0.45
Corg (%)	0.28
F (%)	1.27
CO_2	20.2
LOI	20.7
Cd (mg/kg)	65
Cu (mg/kg)	31
Zn (mg/kg)	215
Mn (mg/kg)	296
Ni (mg/kg)	80

3.3.2. Granulo-Chemical Analysis

The chemical characteristics of the different size fractions of the phosphate sample are shown in Table 3. Results show that the +125–500 µm and the −40 µm fractions (24.3% and 38.8% by weight, respectively) bear the highest amounts in CFA (21.5–11.3% and 13.2% P_2O_5, respectively). All the fractions analyzed are rich in carbonate minerals (CaO concentrations vary from 35.5 to 50.7%). MgO is concentrated in the coarse fraction (+500 µm) and particles of the −125 µm. The fraction (−40 µm) is richer in SiO_2, Fe_2O_3, Al_2O_3 and K_2O than all others (Table 3), which shows that clay minerals are, besides carbonates, among the dominant components of the fine fraction, and that most of SiO_2 is bound to clay minerals. Due to these results, scrubbing and attrition processes may be used in order to remove the major part of slimes and clay coatings, as well as to upgrade the phosphate ore. The trace element concentrations of the different size fractions are presented in Table 4. Results showed that the fine fraction (−40 µm) is richer in Cu and Zn than the others. Cd is concentrated mainly in the coarse fraction (+200 µm).

Table 3. Chemical results of major elements in different size fractions of the studied raw phosphate.

Fraction (µm)	Weight (%)	P_2O_5 (%)	CaO (%)	CaO/P_2O_5	MgO (%)	SiO_2 (%)	SO_3 (%)	Fe_2O_3 (%)	Al_2O_3 (%)	Na_2O (%)	K_2O (%)	Corg (%)	CO_2 (%)	F (%)	LOI
10,000	4.64	4.25	47.3	11.1	3.93	17.0	0.57	0.43	1.10	0.18	0.10	0.10	23.4	1.74	25.0
8000	2.11	4.63	45.1	9.74	4.55	15.6	0.54	0.49	1.24	0.17	0.11	0.12	25.8	1.74	27.5
5000	2.50	5.16	44.2	8.56	3.99	18.5	0.61	0.41	1.20	0.19	0.11	0.34	23.3	2.24	25.4
4000	0.78	5.53	46.2	8.36	3.67	16.6	0.67	0.47	1.20	0.19	0.10	0.04	23.7	2.04	25.2
3150	0.89	6.89	44.2	6.42	3.34	18.0	0.75	0.36	1.14	0.21	0.10	0.24	22.7	2.60	24.59
2000	1.25	7.95	45.4	5.71	3.17	18.2	0.86	0.38	1.10	0.23	0.10	0.21	20.2	1.20	22.1
1000	1.92	8.83	45.7	5.17	3.15	17.9	1.23	0.28	1.09	0.23	0.10	0.24	19.1	2.40	21.0
800	1.06	8.42	46.4	5.51	2.49	17.7	1.09	0.29	1.11	0.22	0.10	0.10	20.2	3.36	21.9
630	1.25	8.27	46.4	5.61	1.35	17.9	1.10	0.34	1.10	0.24	0.12	0.38	20.9	3.44	23.9
500	2.02	9.56	46.8	4.89	3.13	17.1	1.14	0.35	1.09	0.39	0.11	0.37	17.4	3.12	19.5
315	6.05	16.8	49.6	2.94	2.21	16.3	1.67	0.41	1.03	0.57	0.10	0.42	9.09	2.72	11.3
200	9.99	21.5	50.7	2.36	1.53	11.3	2.03	0.29	0.95	0.73	0.09	0.49	8.56	2.44	10.9
160	2.31	18.9	49.8	2.64	2.06	16.5	1.8	0.19	0.94	0.59	0.09	0.16	7.38	2.00	9.16
125	3.94	11.3	45.5	4.04	3.66	13.8	1.13	0.26	0.91	0.43	0.11	0.24	20.9	3.32	22.8
100	3.75	6.86	39.9	5.81	5.25	16.3	0.79	0.24	0.91	0.33	0.11	0.12	26.6	2.52	28.3
90	1.92	6.26	36.6	5.85	4.56	19.6	0.73	0.19	1.10	0.32	0.12	0.19	28.3	0.73	30.1
71	7.11	6.11	35.8	5.86	6.82	21.3	0.66	0.15	0.96	0.22	0.13	0.30	25.4	0.66	27.4
50	5.77	5.26	37.8	7.19	7.21	16.9	0.63	0.20	0.82	0.25	0.08	0.48	28.5	0.63	30.9
40	1.92	5.43	38.5	7.09	7.19	15.8	0.68	0.24	0.86	0.20	0.08	0.12	29.2	0.68	30.9
−40	38.8	13.2	35.5	2.68	5.62	22.5	1.33	0.85	3.35	0.42	1.07	0.28	15.1	1.33	17.1

Table 4. Concentrations (mg/kg) of metallic trace elements in different size fractions of the studied raw phosphate.

Fraction (μm)	Cd	Cu	Zn	Mn	Ni
10,000	95	31	148	574	77
8000	129	27	157	589	134
5000	102	26	138	486	97
4000	108	30	151	492	127
3150	82	24	146	544	106
2000	93	22	151	530	143
1000	98	19	113	522	88
800	104	22	112	575	144
630	110	21	112	613	138
500	103	27	120	621	142
315	106	27	110	439	133
200	80	22	85	340	93
160	67	27	84	378	21
125	76	23	95	418	58
100	57	19	91	296	116
90	47	20	84	253	66
71	40	15	80	220	70
50	40	15	79	211	68
40	45	17	83	236	63
−40	49	49	345	173	87

3.4. Enrichment Methods

As described in the material and methods section, only the +71 μm size fraction was processed by different methods in order to obtain a phosphate-rich fraction (71–315 μm) that is suitable for the flotation process. It is worth noting that the 40–71 μm fraction was subjected to direct and reverse flotation tests in order to assess the flotation performance of phosphate minerals in the fraction below 71 μm. The reverse-flotation test (using the same flowsheet used for the 71–315 μm fractions) was not successful and no stable foam was observed. Direct flotation (2 min) of phosphate minerals using a mixture of 1000 g/t of Aero 725 and fuel oil (pH = 9; 65% solid) gave a low-grade concentrate assaying 15.1% P_2O_5 with a recovery rate of 76.8%. Pease et al. [47] investigated the flotation performance of different size fractions and reported the poor floatability of fine material. Therefore, adding the 40–71 μm fraction to the flotation feed (71–315 μm) might not be a good option.

3.4.1. Scrubbing and Attrition

Scrubbing-attrition treatment was used to improve the grade and recovery of phosphate from the P_2O_5 rich fraction (71–315 μm). The results of the scrubbing test (Table 5) showed that the P_2O_5 grade and recovery are 13.5% and 34.3%, respectively. The P_2O_5 content increased by 1.45%, compared to the representative sample (12.0%), even though this did not have a significant impact on the CaO/P_2O_5 ratio (3.26), compared with that of the raw sample (3.39). MgO and SiO_2 contents decreased from 4.67% to 3.8% and from 20.0% to 16.5%, respectively. Cd concentration also decreased from 65 mg/kg (raw sample) to 55 mg/kg. The scrubbing step has raised the content of P_2O_5 by the elimination of the impurities such as MgO, SiO_2 and Cd. The coarse fraction (+315 μm) represented 10.1% and 19.7% of the feed and phosphate value respectively. This fraction may be ball-milled and combined with the −315 + 71μm fraction to constitute the feed to the flotation circuit. The fine fraction (−71 μm) assays 11.5% of P_2O_5 and a recovery close to 45.9%. Attrition was used to improve the separation efficiency of clay coatings from phosphate surface. Results of the scrubbing-attrition test are shown in Table 6. The 10 min-assay leads to a P_2O_5 grade (15.5%) which is better than those obtained with 5 and 15 min assays (13.7% and 14.4%, respectively).

Table 5. Chemical analysis of scrubbed sample.

Fraction (μm)	Weight (%)	P_2O_5 (%) Grade	P_2O_5 (%) Recovery	CaO (%)	CaO/P_2O_5	MgO (%)	SiO_2 (%)	Cd (mg/kg)
+315	23.1	10.1	19.7	46.8	4.66	3.19	18.8	97
−315+71	30.0	13.5	34.3	43.8	3.26	3.80	16.5	55
−71	46.9	11.5	46.0	35.9	3.11	5.95	22.9	56
Reconstituted	100	11.8	100	40.78	3.46	4.67	20.0	65.2

Table 6. Chemical analysis after scrubbing and attrition treatments.

Time (min)	Fraction (μm)	Weight (%)	P_2O_5 (%) Grade	P_2O_5 (%) Recovery
5	−315+71	88.8	13.7	90.6
10	−315+71	84.2	15.5	97.2
15	−315+71	86.6	14.4	92.4

The scrubbed-attrited 71–315 μm fraction was used for HLS tests, in order to assess the "pure phase" of phosphate, which corresponds to the optimum concentration of P_2O_5 that can be achieved through enrichment methods. Results of HLS and identification of phosphate grains, under the binocular microscope, are shown in Table 7. The highest grade of P_2O_5 (30.9%) achieved by HLS, is the maximum concentration that can be obtained by physical or physicochemical separation processes. CaO is high (49.5%), showing that the carbonates are included into the phosphate particles (endogangue), which is in agreement with the petrographic observations. A high content of CaO in the phosphate concentrate may increase the consumption of sulfuric acid during the production of phosphoric acid [48]. Cd concentration (27 mg/kg) is significantly lower than that of the representative raw sample (65 mg/kg) and the scrubbed one (55 mg/kg). Therefore, it can be stated that cadmium is more concentrated in the exogangue of the studied phosphates.

Table 7. Results of the chemical analysis of pure phosphate phase recovered by HLS.

Element	P_2O_5 (%)	CaO (%)	CaO/P_2O_5	MgO (%)	SiO_2 (%)	Cd (mg/kg)
Concentration	30.9	49.5	1.60	1.09	2.22	27

3.4.2. Grinding

According to Anglaret et al. [38], the rotation speed should be between 65 and 80% of the critical rotational speed (Vcr) of the mill. The critical rotational speed is calculated on the basis of the equation:

$$V_{cr} = 32/\sqrt{D} \quad (1)$$

where D is the diameter of the grinding cell expressed in meters. The diameter of the mill, used in this study, is equal to 0.25 m, which corresponds to a critical rotational speed of 64 rpm; the rotation speed V varies according to the following range: 41.6 < V < 51.2. The rotation speed chosen for this study is 50 rpm.

Effect of the Grinding Time on the Grade and Recovery of P_2O_5

Grinding tests were evaluated at 2, 3 and 4 min after fixing the rotation speed at 50 rpm, solid concentration at 40% and charge ratio at 5/1. Results (Table 8) have shown that the best outcome for P_2O_5 grade and recovery was obtained after a 4 min-grinding period. This optimal grinding time was chosen for the rest of the tests.

Effect of the Charge Ratio on the Grade and Recovery of P_2O_5

The following charge ratio was studied: 3/1, 4/1 and 5/1 after fixing rotation speed at 50 rpm, solid concentration at 40% and grinding time at 4 min. The results displayed in Table 8 show that the best outcome for P_2O_5 grade and recovery was obtained by using the charge ratio 5/1. This optimal ratio was chosen for the rest of the tests.

Effect of Solid Concentration on the Grade and Recovery of P_2O_5

Grinding tests with 30%, 40% and 50% solids were evaluated after fixing the rotation speed at 50 rpm, charge ratio at 5/1 and grinding time at 4 min. Results, displayed in Table 8, show that the best results in term of grade and recovery of P_2O_5 were obtained for 40% solids.

Optimum Grinding Parameters

The optimum grinding parameters are rotation speed (Rs) of 50 rpm, charge ratio (Cr) of 5/1, solid concentration (Sc) of 40% and a grinding time (Gt) equal to 4 min. These parameters allowed us to improve the separation of phosphate grains from their gangue. The optimal test has shown (Table 9) an improvement in P_2O_5 content, from 10.1% (before grinding) to 14.5%. The 71–315 µm fraction released after grinding was subjected to attrition treatment by using the same parameters as the ones employed for the scrubbed fraction (see Section 3.4.1). This step has made it possible to improve the P_2O_5 content from 14.5% to 15.2% (Table 9). Grinding causes modifications in terms of grain texture, morphology, surface chemistry, and electrical-surface charge [49]. The increase of the grain-surface area may increase the reactivity with the flotation reagents [50].

Table 8. Grinding tests results after variation of different parameters.

Parameter to Be Varied	Test	Time (min)	Cr	Sc (%)	Rotation Speed (Rpm)	Weight (%)	P_2O_5 (%)	
							Grade	Recovery
Grinding time (Gt)	1	2	5/1	40	50	28.1	14.2	39.7
	2	3	5/1	40	50	32.9	14.4	47.0
	3	4	5/1	40	50	40	14.5	57.6
Charge ratio (Cr)	1	4	3/1	40	50	29.8	14.4	42.6
	2	4	4/1	40	50	30	14.4	43.1
	3	4	5/1	40	50	40.2	14.5	58.1
Solid concentration (Sc)	1	4	5/1	30	50	28.6	14.4	40.9
	2	4	5/1	40	50	40.2	14.5	58.0
	3	4	5/1	50	50	31.2	14.4	44.7

Table 9. Grade and recovery of P_2O_5 after optimum grinding and attrition tests.

Experiment	Fraction (µm)	Weight (%)	P_2O_5 (%)		CaO (%)
			Grade	Recovery	
Grinding	>315	33.7	6.71	22.5	47.8
	71–315	40.2	14.5	58.0	46.5
	<71	26.1	5.45	14.1	44.8
Attrition	71–315	87.2	15.2	91.4	46.8

3.4.3. Flotation

Flotation tests using anionic/cationic collectors [26,51,52] were applied on the 71–315 µm fractions resulting from sizing, washing, grinding and attrition treatments (Figure 4). The flotation feed contains (Table 10) 15.5% of P_2O_5 and 3.39% MgO with a very high CaO/P_2O_5 ratio (2.9). SiO_2 content (13.2%) is low compared to the representative raw phosphate sample (20.5%) since silicates are preferably associated to the fine fraction. The latter has been reduced by sieving, scrubbing and attrition. During the flotation tests, two parameters were varied: the pH of the pulp and reagents. The siliceous gangue floated at a natural pH by means of a cationic collector [14]. Carbonates are recovered at acidic pH with the anionic collector which operates, at the same time, as an activator of silicates [39]. Phosphate makes the non-floating fraction, which is recovered in acidic pH [31], and in the presence of phosphoric acid, which is the most recommended depressant during calcite flotation [34,53,54]. The content of the major elements in the final products of the optimal flotation test is summarized in Table 10. Tailings are removed as a froth 1 (21.3% weight: silicates > carbonates) and a froth 2 (25.3% weight: carbonates > silicates). The CaO/P_2O_5 ratio is still high (1.71) compared to the ratio required for industrial application (CaO/P_2O_5 < 1.6; [44]). Initially, such a ratio could be attributable to the presence of carbonates included in the phosphate grains, a hypothesis confirmed by Zidi et al. [26] in the study of the low-grade phosphate discharged by the Kef Eddour washing plant. Nevertheless, part of the floated carbonates could result from the similarity of their surface chemical propriety with respect to apatite, which makes their separation rather difficult [33,53]. The contents of SiO_2 (2.52%) and MgO (1.03%) show a noticeable decrease compared to the flotation feed (13.2% and 3.39%, respectively). The valuable ore (53.4% weight) is recovered as non-floating phosphate concentrate assaying 27.1% P_2O_5 with a recovery rate of 92.4%.

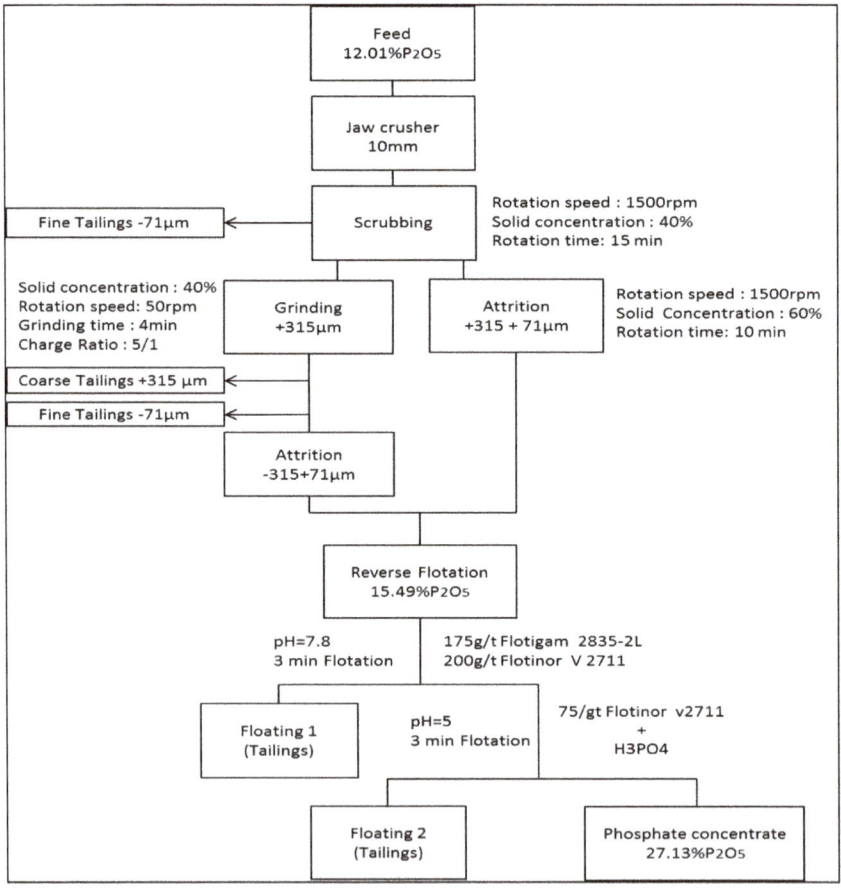

Figure 4. Proposed flow sheet for the valorization of the studied low-grade phosphates.

Table 10. Results of chemical analysis of the flotation feed and the flotation test products.

Product	Weight (%)	P$_2$O$_5$ (%) Grade	P$_2$O$_5$ (%) Recovery	CaO (%)	CaO/P$_2$O$_5$	MgO (%)	SiO$_2$ (%)
Flotation feed	-	15.5	-	44.6	2.9	3.39	13.2
Floating 1 (Tailings)	21.3	3.23	4.40	36.6	11.3	5.96	26.0
Floating 2 (Tailings)	25.3	2.01	3.20	47.2	23.5	4.89	23.9
Non-floating (Phosphate-concentrate)	53.4	27.1	92.4	46.5	1.71	1.03	2.52
Reconstituted	100	15.7	100	44.6	2.80	3.05	12.9

4. Conclusions

Phosphates of the Tozeur-Nefta roof series have, so far, been rejected as non-profitable material. This study has revealed that this material is a low-grade ore (12.0% P$_2$O$_5$) with siliceous and carbonated gangue. Beneficiation assays (sizing, scrubbing-attrition, grinding and flotation) were conducted on the +71 µm fraction of this material. Scrubbing followed by attrition of the 71–315 µm fractions to improve the grade to 15.5% P$_2$O$_5$. The +315 µm fractions were wet-grounded and sieved to 71–315 µm and floated altogether with the fraction 71–315 µm, as yielded by scrubbing-attrition. Reverse-flotation tests conducted on the phosphate-rich fraction (71–315 µm) have made it possible to enhance the grade from 15.5% to 27.1% P$_2$O$_5$ with a recovery of 92.4%. The results of the enrichment tests that were carried out are considered satisfactory as long as they have allowed the beneficiated poor-raw phosphate material to reach the levels of marketable grades. As a result, we believe that the flow sheet developed at the end of this study can be directly applied on an industrial scale.

Author Contributions: F.S. and H.B.; Methodology, H.B., G.D. and H.T.; Investigation, H.B., G.D. and H.T.; Resources, R.S., N.F., N.C., G.D. and H.T.; Writing-Original Draft Preparation, H.B., G.D. and H.T.; Writing-Review & Editing, F.S. and R.S.; Supervision, F.S., R.S., N.F. and N.C.; Project Administration, F.S.

Funding: This research received no external funding.

Acknowledgments: The authors owe a great appreciation to Compagnie des Phosphates de Gafsa (CPG) for all the facilities granted to the sampling campaign as well as the mineralogical (Haithem Mansouri) and chemical analysis (Sahbi Harakati). The manuscript benefited from thorough critical appraisal by Bousairi Boujlel, former Central Research Director of CPG. In addition, we thank Faysal Souissi, English Department, College of Humanities of Tunis, University of Tunisia, for his kind and thorough review of the English version of this manuscript.

Conflicts of Interest: The authors declare no conflict of interest.

References

1. Galfati, I. Etude de l'impact des rejets de l'industrie Phosphatière sur l'environnement dans le bassin de Gafsa Métlaoui. Ph.D. Thesis, University of Tunis El Manar, Tunis, Tunisia, 2010.
2. IFDC. Available online: https://ifdc.org/wp-content/uploads/2015/03/100922_kauwenbergh_presentation_0.pdf (accessed on 14 September 2018).
3. Burollet, P.F. Contribution à l'étude stratigraphique de la Tunisie centrale. *Ann. Mines Geol.* **1956**, *18*, 1–350.
4. Sassi, S. La sédimentation phosphatée au Paléocène dans le Sud et le Centre oust de la Tunisie. Ph.D. Thesis, Université d'Orsay, Paris-Sud, France, 1974.
5. Chaabani, F. Dynamique de la partie orientale du bassin de Gafsa au crétacé et au paléogène. Etude minéralogique et géochimique de la série phosphatée éocène (Tunisie méridionale). Ph.D. Thesis, University of Tunis El Manar, Tunis, Tunisia, 1995.
6. Zaier, A. Evolution tecto-sédimentaire du bassin phosphate du Centre-Ouest de la Tunisie, minéralogie, pétrographie, géochimie et genèse des phosphorites. Ph.D. Thesis, University of Tunis El Manar, Tunis, Tunisia, 1999.
7. Bel Haj Khalifa, M. Etude géostatistique du gisement de phosphates multicouches de Nefta-Tozeur (Tunisie). Ph.D. Thesis, University of Tunis El Manar, Tunis, Tunisia, 1996.

8. Ben Hassen, A. Données nouvelles sur la matière organique associée aux séries du bassin phosphaté du sud-tunisien (Gisement de Ras-Draâ) et sur la phosphatogenèse. Ph.D. Thesis, Université d'Orléans, Orléans, France, 2007.
9. Ben Hassen, A.; Trichet, J.; Disnar, J.R.; Belayouni, H. Données nouvelles sur le contenu organique des dépôts phosphatés du gisement de Ras-Draâ (Tunisie). *C. R. Geosci.* **2009**, *341*, 319–326. [CrossRef]
10. Ben Hassen, A.; Trichet, J.; Disnar, J.R.; Belayouni, H. Pétrographie et géochimie comparées des pellets phosphatés et de leur gangue dans le gisement phosphaté de Ras-Draâ (Tunisie). Implications sur la genèse des pellets phosphatés. *Swiss J. Geosci.* **2010**, *103*, 457–473. [CrossRef]
11. Gallala, W.; Saïdi, M.; El Haji, S.; Zayani, K.; Gaied, M.E.; Montacer, M. Characterization and Valorization of Tozeur-Nefta Phosphate Ore Deposit (Southwestern Tunisia). *Procedia Eng.* **2016**, *138*, 8–18. [CrossRef]
12. El-Jallad, I.S.; Abouzeid, A.Z.; El-Sinbawy, H.A. Calcination of phosphates: Reactivity of calcined phosphate. *Powder Technol.* **1980**, *26*, 187–197. [CrossRef]
13. Zafar, Z.I.; Anwar, M.M.; Pritchard, D.W. Optimization of thermal beneficiation of a low-grade dolomitic phosphate rock. *Int. J. Miner. Process.* **1995**, *43*, 123–131. [CrossRef]
14. Mohammad Khani, M.; Noaparast, M.; Shafaei, S.Z.; Amini, A.; Amini, E.; Abdollahi, H. Double reverse flotation of a very low-grade sedimentary phosphate rock, rich in carbonate and silicate. *Int. J. Miner. Process.* **2011**, *100*, 157–165. [CrossRef]
15. Good, P.C. *Beneficiation of Unweathered Indian Calcareous Phosphate Rock by Calcination and Hydration*; US Bureau of Mines: Washington, DC, USA, 1976.
16. Hollick, C.T.; Wright, R. Recent trends in phosphate mineral beneficiation. *Trans. Inst. Min. Metall. Sect. A* **1986**, *95*, 150–154.
17. Rao, T.C.; Rao, L.S.; Rao, G.M. Beneficiation of Indian Low-Grade Phosphate Deposits-Problems and Prospects. *Trans. Indian Inst. Met.* **1992**, *45*, 195–205.
18. Van Straaten, P. *Rocks for Crops: Agrominerals of Sub-Saharan Africa*; ICRAF: Nairobi, Kenya, 2002; ISBN 0-88955-512-5.
19. Woodrooffe, H.M. Phosphate in the Kola Peninsula, USSR. *Miner. Eng.* **1972**, *24*, 54–56.
20. Blazy, P.; Jdid, E.A. Calcination du phosphate sédimentaire à gangue carbonatée d'Akashat (Irak) en four rotatif et en four éclair (Flash). *C. R. Acad. Sci* **1997**, *325*, 761–764. [CrossRef]
21. Bangar, K.C.; Yadav, K.S.; Mishra, M.M. Transformation of rock phosphate during composting and the effect of humic acid. *Plant Soil* **1985**, *25*, 259–266. [CrossRef]
22. Sadeddin, W.; Abu-Eishah, S.I. Minimization of free calcium carbonate in hard and medium-hard phosphate rocks using dilute acetic acid solution. *Int. J. Miner. Process.* **1990**, *30*, 113–125. [CrossRef]
23. Abu-Eishah, S.I.; Muthaker, M.; Touqan, N. A new technique for the beneficiation of low-grade carbonate-rich phosphate rocks by digestion with dilute acetic acid solutions: Pilot plant testing results. *Miner. Eng.* **1991**, *4*, 573–586. [CrossRef]
24. Ibrahim, S.S.; El Kammar, A.M.; Guda, A. Characterization and separation of pyrite from Abu Tartur black shale. *Int. J. Min. Sci. Technol.* **2015**, *25*, 565–571. [CrossRef]
25. Mâamri, A.J.; Abbassi, L.; Batis, H.N. Characterization of the Oum El Khacheb phosphorites (South Tunisia) and enrichment of big rejections by grinding. *Int. J. Min. Sci. Technol.* **2016**, *26*, 833–842. [CrossRef]
26. Zidi, R.; Babbou-Abdelmalek, C.; Chaabani, F.; Abbassi, L. Enrichment of low-grade phosphate coarse particles by froth-flotation process, at the Kef-Eddour washing plant, Tunisia. *Arab. J. Geosci.* **2016**, *9*, 462. [CrossRef]
27. Lawver, J.E.; Weigel, R.L.; Snow, R.E.; Hwang, C.L. Phosphate reserves enhanced by beneficiation. *Min. Congr.* **1982**, *68*, 27–31.
28. Elgillani, D.A.; Abouzeid, A.Z. Flotation of carbonates from phosphate ores in acidic media. *Int. J. Miner. Process.* **1993**, *38*, 235–256. [CrossRef]
29. Abouzeid, A.Z.M. Physical and thermal treatment of phosphate ores-an overview. *Int. J. Miner. Process.* **2008**, *85*, 59–84. [CrossRef]
30. El-Midany, A.A. Separating dolomite from phosphate rock by reactive flotation: Fundamentals and application. Ph.D. Thesis, University of Florida, Gainesville, FL, USA, 2004.
31. Henchiri, A.; Cecile, J.L.; Baudet, G.; Barbery, G.; Bloise, R.U.S. Process of the treatment of phosphate ores with silico-carbonate gangue. U.S. Patent 4,324,653, 13 April 1982.
32. Houot, R. Beneficiation of phosphatic ores through flotation: Review of industrial applications and potential developments. *Int. J. Miner. Process.* **1982**, *9*, 353–384. [CrossRef]

33. Abdel-Khalek, N.A. Evaluation of flotation strategies for sedimentary phosphates with siliceous and carbonates gangues. *Miner. Eng.* **2000**, *13*, 789–793. [CrossRef]
34. Abouzeid, A.Z.; Negm, A.T.; Elgillani, D.A. Upgrading of calcareous phosphate ores by flotation: Effect of ore characteristics. *Int. J. Miner. Process.* **2009**, *90*, 81–89. [CrossRef]
35. Lisiansky, L.; Baker, M.; Larmour-Ship, K.; Elyash, O. A Tailor Made Approach for the Beneficiation of Phosphate Rock. In *Beneficiation of Phosphates*; Zhang, P., Miller, J.B., Wingate, E., Filho, L.L., Eds.; SME CO.: Englewood, CO, USA, 2016; pp. 55–61. ISBN 978-0-87335-427-1.
36. Liu, X.; Zhang, Y.; Liu, T.; Cai, Z.; Sun, K. Characterization and Separation Studies of a Fine Sedimentary Phosphate Ore Slime. *Minerals* **2017**, *7*, 94. [CrossRef]
37. Somasundaran, P.; Lin, I.J. Effect of the Nature of Environment on Comminution Processes. *Ind. Eng. Chem. Process. Des. Dev.* **1972**, *11*, 321–331. [CrossRef]
38. Anglaret, P.; Filippi, J.; Kazmierczak, S. *Technologie Génie Chimique, Tome I*; CRDM: Amiens, France, 1998; pp. 17–384. ISBN 2-86615-223-9.
39. Houot, R.; Joussemet, R.; Tracez, J.; Brouard, R. Selective flotation of phosphatic ores having a siliceous and/or a carbonated gangue. *Int. J. Miner. Process.* **1985**, *14*, 245–264. [CrossRef]
40. Henchiri, A. A contribution to carbonates–phosphate separation by flotation technique. In *Beneficiation of Phosphate: Theory and Practice*; El-Shall, H., Moudgil, R., Wiegel, R., Eds.; SME: Littleton, CO, USA, 1993; pp. 225–243.
41. Baudet, G.; Save, M. Phosphoric esters as carbonate collectors in the flotation of sedimentary phosphate ores. In *Beneficiation of Phosphates: Advances in Research and Practice*; Zhang, P., El-Shall, H., Wiegel, R., Eds.; SME: Littleton, CO, USA, 1999; pp. 163–185.
42. Blazy, P. *La valorisation des Minerais: Manuels de Minéralurgie*; Presses Universitaires de France: Vendôme, France, 1970; pp. 5–415.
43. Sengul, H.; Ozer, A.K.; Gulaboglu, M.S. Beneficiation of Mardin-Mazıdaği (Turkey) calcareous phosphate rock using dilute acetic acid solutions. *Chem. Eng. J.* **2006**, *122*, 135–140. [CrossRef]
44. Becker, P. *Phosphates and Phosphoric Acid: Raw Materials, Technology, and Economics of the Wet Process*; Marcel Dekker: New York, NY, USA, 1989; pp. 1–579. ISBN 0-8247-1712-0.
45. Khelifi, L. Contribution à l'étude géochimique des phosphates du bassin de Gafsa-Metlaoui. Exemple du gisement d'Oum Lakhcheb. Master's Thesis, University of Tunis El Manar, Tunis, Tunisia, 2012.
46. Galfati, I.; Bilal, E.; Abdallah, H.; Beji-Sassi, A. Geochemistry of solid effluents and phosphate ore washed from Metlaoui-Gafsa basin, Tunisia. *Rom. J. Miner. Depos.* **2014**, *87*, 83–86.
47. Pease, J.D.; Curry, D.C.; Young, M.F. Designing flotation circuits for high fines recovery. *Miner. Eng.* **2016**, *19*, 831–840. [CrossRef]
48. Al-Fariss, T.F.; El-Aleem, F.A.A.; El-Nagdy, K.A. Beneficiation of Saudi phosphate ores by column flotation technology. *J. King Saud Univ. Sci* **2013**, *25*, 113–117. [CrossRef]
49. Clerc, L. Broyage ultrafin de carbonates naturels: Paramétrisation, modélisation et conséquences physico-chimiques. Ph.D. Thesis, Ecole Nationale Supérieure des Mines, Saint-Etienne, France, 1983.
50. Bafghi, M.S.; Emami, A.H.; Zakeri, A.; Khak, J.V. Development and verification of a mathematical model for variations of the specific surface area of mineral powders during intensive milling. *Powder Technol.* **2010**, *197*, 87–90. [CrossRef]
51. Jian, T.; Longhua, X.; Wei, D.; Hao, J.; Zhiyong, G.; Yuehua, H. Adsorption mechanism of new mixed anionic-cationic collectors in a spodumene-Feldspar flotation system. *Chem. Eng. Sci.* **2017**, *164*, 99–107. [CrossRef]
52. Houqin, W.; Jia, T.; Longhua, X.; Shuai, F.; Zhenye, Z.; Ruan, C. Flotation and adsorption of new mixed anionic/cationic collector in the spodumene-feldspar system. *Miner. Eng.* **2018**, *127*, 42–47. [CrossRef]
53. Amankonah, J.O.; Somasundaran, P. Effects of dissolved mineral species on the electrokinetic behavior of calcite and apatite. *Colloids Surf.* **1985**, *15*, 335–353. [CrossRef]
54. Hsieh, S.S.; Lehr, J.R. Beneficiation of dolomitic Idaho phosphate rock by the TVA diphosphoric acid depressant process. *Min. Metall. Explor.* **1985**, *12*, 10–13. [CrossRef]

© 2018 by the authors. Licensee MDPI, Basel, Switzerland. This article is an open access article distributed under the terms and conditions of the Creative Commons Attribution (CC BY) license (http://creativecommons.org/licenses/by/4.0/).

Article

A Comparative Study of Different Columns Sizes for Ultrafine Apatite Flotation

Elves Matiolo [1,*], Hudson Jean Bianquini Couto [1], Michelle Fernanda de Lira Teixeira [2], Renata Nigri de Almeida [1] and Amanda Soares de Freitas [1,3]

1. Centro de Tecnologia Mineral (CETEM/MCTI), Rio de Janeiro 21941-908, Brazil
2. Copebras—China Molybdenum Company Ltd (CMOC International Brasil), Ouvidor (GO) 75715-000, Brazil
3. Centro de Tecnologia Mineral (CETEM/MCTIC), Universidade Federal do Rio Grande do Sul (UFRGS), Porto Alegre 90040-060, Brazil
* Correspondence: ematiolo@cetem.gov.br; Tel.: +55-21-38657334

Received: 19 April 2019; Accepted: 21 June 2019; Published: 28 June 2019

Abstract: The desliming operation to discharge ultrafine particles less than 20 µm prior to concentration by flotation is a common practice in phosphate ores beneficiation plants. The first industrial application for the beneficiation of the phosphate material with particle sizes <44 µm in Brazil was in the Araxá plant concentrator in the beginning of the 1980s. This work shows the comparative flotation results with two different phosphate slime samples (<40 µm) obtained from the Copebras (CMOC International) industrial plant located in Catalão (Goiás state, Brazil), considering a circuit with rougher/cleaner configuration with different columns sizes, as follows: Circuit 1 (rougher—4″ diameter column; cleaner—2″ diameter column) and circuit 2 (rougher—6″ diameter column; cleaner 4″ column). The results indicate that better flotation apatite recovery results were achieved for the circuit with higher size columns (6″ and 4″). The results can be explained by the application of a cavitation tube in the rougher stage in the 6″ column. The improved flotation performance can be attributed to increased probabilities of collision and attachment and the reduced probability of detachment by the small size bubbles generated by the cavitation tube in comparison with the bubbles produced by the porous tube of the 4″ column flotation.

Keywords: apatite; flotation; column flotation; slimes

1. Introduction

Flotation is considered the most effective technology to concentrate phosphate bearing minerals from sedimentary and igneous phosphates ores [1–3]. One of the main disadvantages of phosphate upgrading by flotation is the high sensitivity of collectors to slimes (size fraction <44 µm). Therefore, desliming is a necessary pre-request for successful flotation [4–7] and, following this trend, the classical aim of desliming is to remove fine particles (<37–45 µm) from the flotation feed. Fine/ultrafine particles account for 10–30% of losses of phosphate values in the worldwide beneficiation of phosphate ores, for instance, in Florida [8], India [9], Iran [3], Australia [6], Jordan [10], China [11], and Brazil [12–15].

There are several characterization and process studies focused on the exploitation of phosphate bearing minerals from slimes. In the case of slimes from phosphate ores from Florida, chemical and physical characterizations of slimes show that the P_2O_5 content ranges from 10% to 25% and the particle size distribution indicates that d_{80} are under 30 µm with significant amount (>10–30%) under 2–5 µm. Additionally, the mineralogical composition of the slimes indicates a high content of clay minerals, which are extremely deleterious to the flotation process [8]. Similar chemical and physical characteristics of the slimes can be found in phosphate ores from Brazil [13–15], Australia [6], and India [9]. Regarding the technological approaches to the concentration of ultrafine phosphate bearing minerals from slimes, one strategy is based on desliming in hydrocyclones with low diameters (40 mm) to remove particles under 5–10 µm, followed by the flotation of the desliming product [8,13–15]. Other alternatives tested as well were flotation using a Jameson Cell [6], column flotation [3,16], and also flotation with nanobubble injection in conventional flotation machines without desliming prior to flotation [3].

Pradip and Sankar [9] carried out studies with a slime sample from the Maton rock phosphate beneficiation plant located in Rajasthan, India. One of the major sources of losses in phosphate values is due to slimes generated during the washing of the crushed ore product, called "crusher slimes". The slimes fraction (75%; 37 µm) consists of carbonate-fluorapatite in association with quartz, kaolinite, and limonite, and assays 18–21% P_2O_5, accounting for 10–15% losses of phosphate values in the plant. In this work, the authors concluded that it is possible to float phosphate ore slimes, all passing 37 µm (50%; 8 µm), from Maton, India using a sodium oleate collector, under reagent conditions very different from what is required and currently practiced in the plant for a relatively coarser feed (80%; 74 µm). It should also be possible to condition the slimes separately and then combine them with the coarse ore feed in the existing flotation circuit in the plant.

In contrast with the established methods of phosphate beneficiation where classification by hydrocyclone is mainly used to remove ultrafine particles as tailings, Teague and Lollback [6] describe a process which successfully beneficiates phosphate ore with a high proportion of ultrafine particles (80% <20 µm) to be recovered via flotation without prior classification to remove ultrafine particles (<2–5 µm). The process uses conditioning with reagents at high wt % solids (at least 70 wt %) and flotation with Jameson cells in a rougher, scavenger, cleaner configuration to recover at least 80% P_2O_5 at a grade of 32% P_2O_5 or greater. The Jameson cell was found to have an advantage over conventional flotation cells when treating ultrafine particles, due to their intense mixing zone and propensity to form small bubbles. Pourkarimi et al. [3] carried out studies with a slime sample from the Esfordi phosphate processing plant from the central desert in Iran in Yazd province. The ore contained about 10% of the mass, with a grade of more than 16% P_2O_5 and d_{80} of less than 30 µm. The authors compared the results considering flotation in the presence and absence of nanobubbles using a Denver flotation cell and carried out the generation of nanobubbles based on the cavitation phenomenon through a venturi tube. The flotation results indicated that the existence of nanobubbles in the process had positive effects, so that more than 90% of phosphates with a grade of more than 40% were recoverable in the presence of nanobubbles, while in their absence, under the same conditions, the P_2O_5 grade was 37%. The recovery of apatite in flotation in the presence of nanobubbles significantly increased up to more than 30%, compared to flotation in the absence of nanobubbles.

One alternative to the flotation of phosphate slimes is the application of the column flotation [10,12–16]. Al-Thyabat et al. [7] evaluated the floatability of Jordanian phosphate slime (<38 µm) in a batch column flotation cell, 100 cm high and 5 cm in diameter. Flotation was conducted using sodium oleate as a collector and methyl isobutyl carbinol (MIBC) as a frother. Rougher flotation yielded a concentrate assaying 29.19% P_2O_5 with 90.29% recovery at the following flotation parameters: A total 2.6 cm/s superficial gas velocity, 20 ppm frother dosage, 0.205 mmol/L collector concentration, and 40% weight solids. It was also found that conditioning with the same collector dosage, but with a different solids weight percent, gave different flotation results.

The Brazilian phosphate industry has made great efforts to develop technology for the production of phosphates from slimes as shown by Guimarães and Peres [12–14]. The first industrial application was in the Araxá concentrator in the beginning of the 1980s, in a circuit consisting of desliming in 40 mm hydrocyclones and apatite flotation in column machines. This technology has been upgraded since 1984 by means of significant improvements in the desliming operation, concerning equipment and process control, utilization of a more adequate reagent system, and the use of flotation machines that provide improved performance and metallurgical yield. The concept of this process was still expanded to other Brazilian plants located in Catalão (Goiás state), Tapira (Minas Gerais state), and Cajati (São Paulo state) in the 1990s. It is estimated that the apatite concentrate from slimes represents 11% to 13% of the overall production. The apatite concentrate grade achieved from slimes are P_2O_5

at 33.0–35%; Fe$_2$O$_3$ at 3.0–6.5%; Al$_2$O$_3$ at 0.4–1.0%; MgO at 0.1–1.6%, and SiO$_2$ at 1.5–5.5%. More recently, Matiolo et al. [15] showed the results of a flowsheet developed for apatite concentration from a slime sample provided by the industrial phosphate ore plant Copebras (China Molybdenum-CMOC International), located in Catalão (Goiás state, Brazil). Considering the flowsheet with the desliming in the hydrocyclones (40 mm in diameter) in two stages, followed by apatite flotation in the column, a final concentrate of 35.6% P$_2$O$_5$ was obtained, with main impurities of Fe$_2$O$_3$ and SiO$_2$ around 4.0% and 5.1%, respectively, using only rougher flotation. Taking into account the mass and metallurgical balances of desliming and flotation, the overall mass recovery was around 9.5% and the P$_2$O$_5$ recovery was 25.5%.

This work is the continuation of previous work with the Copebras phosphate slime samples. by Matiolo et al [15] This article does a comparison considering a circuit with rougher/cleaner configuration with different sized columns flotations, as follows: Circuit 1 (ro—ugher4" diameter column; cleaner—2" diameter column), and circuit 2 (rougher—6" diameter column; cleaner 4" column). The influence of collector and depressant dosages, pulp pH, superficial wash water velocity (Jw), ore variability, and superficial air velocity (Jg) were evaluated on the flotation separation parameters (P$_2$O$_5$ grade and recovery and impurities content) for both circuits. The chemical quality target for the apatite concentrate is P$_2$O$_5$ > 33.0%, SiO$_2$ < 8%, and Fe$_2$O$_3$ < 6.5%.

2. Experimental

2.1. Ore Sample

The two slime samples tested in this work were provided by Copebras (CMOC International) located in Catalão (Goiás state), Brazil. Detailed information on the sample preparation procedures can be found in the study by Matiolo et al. [15]. The slime samples were collected from the overflow of the hydrocyclones, which was around 10 tons (dry basis) of the slime in pulp with 16% solid content by weight. The sample characterization included chemical analyses by X-ray fluorescence (PANalytical AXIOS WDS, PANalytical, Almero, the Netherlands) and mineralogical characterization by XDR (Rietveld method) (Bruker-D4 Endeavor, Bruker, Germany). The particle size distribution was determined by the laser diffraction technique using a Malvern Master size particle analyzer (Malvern Panalytical Ltd., Malvern, UK).

The particle size distribution of slime Sample 1 showed that the characteristics for diameters D_{32}, D_{10}, D_{50}, and D_{90} were, respectively, 2.6 µm, 1.0 µm, 5.7 µm, and 24.7 µm, whereas for Sample 2 the values were D_{32} = 5.0 µm, D_{10} = 2.1 µm, D_{50} = 13.8 µm, and D_{90} = 43 µm. The chemical analyses for both samples showed that the P$_2$O$_5$ grade was around 12–13% and the CaO was around 11–14%. In terms of the major contaminants, SiO$_2$ and Fe$_2$O$_3$ stood out, with grades of 19.7% and 27.7% for Sample 1 and 15.8% and 24.8% for Sample 2, respectively (Table 1). The main mineral phases in the samples were apatite (around 33%), goethite with a 26% grade content, and quartz with 18%.

Table 1. Chemical analysis of the slime samples.

Sample	Al$_2$O$_3$	BaO	CaO	Fe$_2$O$_3$	MgO	P$_2$O$_5$	SiO$_2$	Nb$_2$O$_5$	CaO/P$_2$O$_5$
1	3.9	1.2	14.0	27.7	2.4	13.3	19.7	0.62	1.05
2	2.8	1.6	16.5	30.0	1.0	14.0	12.5	0.89	1.18

2.2. Desliming

Figure 1 shows the apparatus for the desliming operation of the slime samples. Each container, holding 1 m^3 (40 in total) of the slime sample, was connected to a vertical pump with a system that allowed pulp recirculation to the container and also to a 3.6 m^3 tank. In this tank, water was added to adjust the solids in the pulp to around 8%, which represented the solids to the feed of the desliming stage. From the storage tank, the pulp was pumped to the hydrocyclone apparatus. Desliming was carried out in two stages, where the underflow from the first stage fed the second stage. The underflow of the second stage fed the apatite flotation and the overflow from both stages was discharged as tailings. The hydrocyclone used was supplied by Weir Minerals. The apex finder was 7 mm in the first stage and 5.5 mm in the second. In both stages, the vortex finder was 10 mm and the operational pressure was 4 kgf/cm^2 in the first stage and 3 kgf/cm^2 in the second.

Figure 1. Apparatus for the desliming operation.

3. Flotation Studies

3.1. Reagents

Soybean oil soap was used as an apatite collector and was obtained after saponification for a period of 15 min with a fatty acid/NaOH ratio of 5:1. Gelatinized corn starch was used as the gangue mineral depressant, with a starch/NaOH ratio of 4:1 and a reaction time of 10 min, in a solution with concentration at 20% *w/w*. After the saponification and gelatinization, both reagents were diluted in distilled water, producing 1% *w/w* solutions. For pH adjustment, a 2% NaOH solution was used. Tap water from the city of Rio de Janeiro's supply network was used to attain the correct percentage of pulp solids in the flotation studies.

3.2. Pilot Plant Flotation Tests

Figure 2a shows the schematic flowsheet considering the desliming and flotation stages for apatite concentration from the slime samples and Figure 2b shows a general view of the columns at CETEM's pilot plant (ERIEZ, Delta, Canada). The underflow obtained in the second stage of desliming, at 35% solid content by weight, fed the flotation circuit. The rougher/cleaner flotation trials were performed using columns with internal diameters of 6", 4", and 2" and a height of 7.0 m. The effective volume of each column was 100 L, 46 L, and 13 L, respectively. The main objective of this study was to compare the flotation performance considering a rougher/cleaner circuit for apatite flotation with the following two different circuits: Circuit 1 (rougher—4" diameter column; cleaner—2" diameter column) and circuit 2 (rougher—6" diameter column; cleaner 4" column). The solid feed rate in each circuit was 35 kg/h and 46 kg/h dry basis, respectively.

From the storage tank, the pulp was pumped to a cylindrical tank where it was conditioned with gelatinized corn starch and the pH was adjusted (between 9.5 and 10.8) with NaOH. The depressant dosage ranged from 2300 g/t up to 3000 g/t. After conditioning with the depressant, the pulp was conditioned with soybean oil soap in a cylindrical tank, at 35% solids by weight, with dosages from 70 g/t to 130 g/t. The mean residence time for conditioning was 10 min for the depressant and around 20 min for the collector. After conditioning, the pulp was diluted to 20% solids by weight and fed in the rougher column. The froth obtained in the rougher fed the cleaner, and the sink fractions from rougher and cleaner were discharged as final tailings. Bubbles were generated with controlled pressure and flow via a forced air passage in a porous tube at the bottom of the 4" and 2" columns. The bubble generation of the 6" column flotation was through the recirculation of a portion of the pulp through a cavitation tube. The operational pressure on the cavitation was around 3.5 kgf/cm^2. The pulp/froth interface was controlled by a level sensor that was connected to the tailings pump and the wash water and air flow rate were controlled by a flow meter. After reaching the stationary stage, samples of the rougher and cleaner tailings and cleaner concentrate were collected simultaneously for 2 min for mass and metallurgical balances. The samples were flocculated and dried in an oven for 24 h at 80 °C. Then, the samples were weighed, disaggregated, and separated in aliquots for chemical analysis by X-ray fluorescence spectroscopy to determine the metallurgical balance.

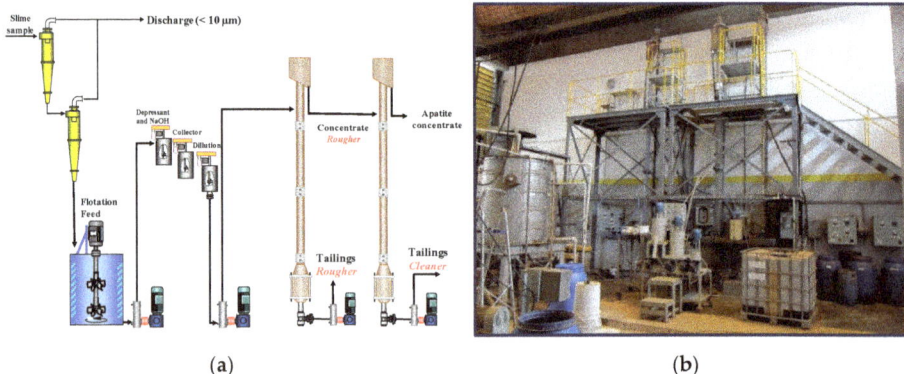

Figure 2. Schematic flowsheet for the apatite concentration from (**a**) the slime samples and photo from (**b**) the column flotation units.

4. Results and Discussion

4.1. Desliming

The mass and oxide balance for the optimized desliming conditions for Sample 1 are shown in Table 2. The feed of the first desliming stage contained 8% solids by weight and produced an underflow with 16% solids by weight and an overflow with 2–3% solids by weight, which was discharged as final tailings. The mass and P$_2$O$_5$ recoveries of the underflow were 70% and 78%, respectively. The P$_2$O$_5$ grade increased from 13.3% at the feed to 14.3% in the underflow. The P$_2$O$_5$ in the overflow was 10.1%. The underflow of the first stage fed the second stage, which was carried out in the same hydrocyclone applied in the first stage with an apex finder of 5.5 mm and the operation pressure was regulated to 3 kgf/cm^2. The solids content in the overflow of the second stage was 4–5%, the d_{50} was 5 µm, and the P$_2$O$_5$ grade was 11% and was discharged with the final tailings together with the overflow produced at the first stage. The solids content in the underflow of the second stage was 35–36%, the d_{50} was 18 µm, and the P$_2$O$_5$ grade was 16.4%. The mass and P$_2$O$_5$ recoveries at the second stage were very similar to the first stage, reaching 70% and 77%. The underflow obtained at the second stage fed the conditioning circuit and then the flotation circuit. Considering both stages, the overall mass recovery was around 50% and the P$_2$O$_5$ recovery was 60%. Regarding the SiO$_2$ and Fe$_2$O$_3$ contaminants, it was observed that the iron impurities tended to decrease after desliming, as opposed to what was observed for the SiO$_2$ content, which tended to increase after this process.

The results of the desliming stage for Sample 2 are shown in Table 3. The P$_2$O$_5$ content at the flotation feed (underflow of second stage) was quite similar to that obtained for Sample 1, assaying around 16%. Regarding the SiO$_2$ and Fe$_2$O$_3$ impurities, a reduction of the iron bearing minerals content in the flotation feed was observed when compared to the feed sample, as observed in Sample 1. The SiO$_2$ content in the flotation feed was almost the same as that in the feed process sample.

Table 2. Metallurgical and mass balance of the desliming stage (Sample 1).

Stage	Stream	Mass Rec (%)	Grade (%)						Distribution (%)					
			Al_2O_3	CaO	Fe_2O_3	MgO	P_2O_5	SiO_2	Al_2O_3	CaO	Fe_2O_3	MgO	P_2O_5	SiO_2
1	Over	29.9	4.5	9.9	33.9	2.3	10.1	16.5	35.4	21.4	36.9	29.0	23.1	25.3
	Under	70.1	3.5	15.5	24.7	2.4	14.3	20.7	64.5	78.5	63.0	70.9	76.8	74.6
2	Over	29.8	4.4	11.5	31.1	2.5	11.3	17.3	37.5	21.3	40.0	31.5	22.5	24.0
	Under	70.2	3.1	17.9	19.7	2.3	16.4	23.2	62.4	78.6	59.9	68.4	77.4	76.0
	Feed	100	3.8	13.8	27.4	2.4	13.0	19.4	100	100	100	100	100	100

Table 3. Metallurgical and mass balance of the desliming stage (Sample 2).

Stage	Stream	Mass Rec (%)	Grade (%)						Distribution (%)					
			Al_2O_3	CaO	Fe_2O_3	MgO	P_2O_5	SiO_2	Al_2O_3	CaO	Fe_2O_3	MgO	P_2O_5	SiO_2
1	Over	43.2	3.7	12.9	35.3	1.2	11.5	12.3	58.4	33.9	51.4	43.1	34.8	42.6
	Under	56.8	2.0	19.2	25.5	1.2	16.4	12.6	41.6	66.1	48.6	56.9	65.2	57.4
2	Over	15.0	2.8	10.9	28.8	1.1	12.7	10.5	20.3	8.9	17.9	13.5	11.7	12.0
	Under	85.0	1.9	19.8	23.4	1.2	16.9	13.3	79.7	91.1	82.1	86.5	88.3	88.0
	Feed	100	2.7	16.5	29.7	1.2	14.3	12.5	100	100	100	100	100	100

4.2. Flotation Studies

This topic presents the results and discussion of the apatite flotation studies considering rougher/cleaner flotation in 6", 4", and 2" columns. As mentioned before, one circuit consisted of the application of a 6" column for the rougher stage and a 4" column for the cleaner. The second circuit applied the 4" column for rougher flotation and a 2" column for the cleaner. The influence of collector and depressant dosages, wash water superficial velocity, pH, and air superficial velocity were evaluated on flotation separation parameters. The flotation recovery was calculated as a function of the hydrocyclone underflow (flotation feed).

Figure 3 shows the curve of the P_2O_5 grade versus the recovery, considering apatite rougher/cleaner configuration for both circuits evaluated. As shown in the figure, the P_2O_5 recovery ranged from 80% to 25% with a P_2O_5 grade variation from 22% to almost 38%. In general terms, the flotation performance (grade and recovery) in the circuit with 6" and 4" columns was better compared to the circuit with 4" and 2" columns, especially for the P_2O_5 grade between 30% and 34%.

Figure 4a shows the comparative results for the P_2O_5 grade/recovery curve considering the rougher flotation in the 6" and 4" columns and for cleaner flotation considering the 4" and 2" columns (Figure 4b). The results for the circuit applied in the 6" and 4" columns were obtained with the flotation of Samples 1 and 2 and considering the Sample 1 flotation for the circuit with the 4" and 2" diameter columns. The performance (grade and recovery) of rougher flotation in the 6" diameter column was superior to the results obtained for the 4" diameter column. As it can be observed, the P_2O_5 recovery ranged from 90% to 40%, with a P_2O_5 grade variation from 17% to almost 34%. Furthermore, a linear relation between the P_2O_5 grade and losses was verified for the rougher tailings.

A fundamental difference between both columns is the fact that the bubble generator of the 6" column is a cavitation tube, while the bubble generator of the 4" column is a porous tube. Hydrodynamic cavitation to generate bubbles has been use to enhanced the flotation of fine and ultrafine particles [17–19]. Tao et al. [17] evaluated the effect of picobubble injection produced by the hydrodynamic cavitation principle in association with the conventional sized bubbles produced by a static mixer on the flotation response of fine coal particles. The results indicated that picobubbles significantly enhanced the coal flotation process with higher recovery and lower product ash. The flotation recovery increased by 10–30% depending on the process operating conditions. Zhou et al. [18] incorporated a cavitation tube in the feed line to a conventional flotation cell for the flotation of fine silica and ZnS precipitates (<5 μm). The results showed that there is a substantial increase in fine silica recovery for a given flotation period when using the cavitation tube (without added air). Additionally, a 40% increase in rate constant was obtained using a cavitation tube (1.3 mm nozzle diameter), even though the overall aeration was less (2.15 L/min compared to 3 L/min without the tube). This increase in the flotation rate constant again suggests that small bubbles generated by cavitation in the feed stream played a role in enhancing flotation kinetics.

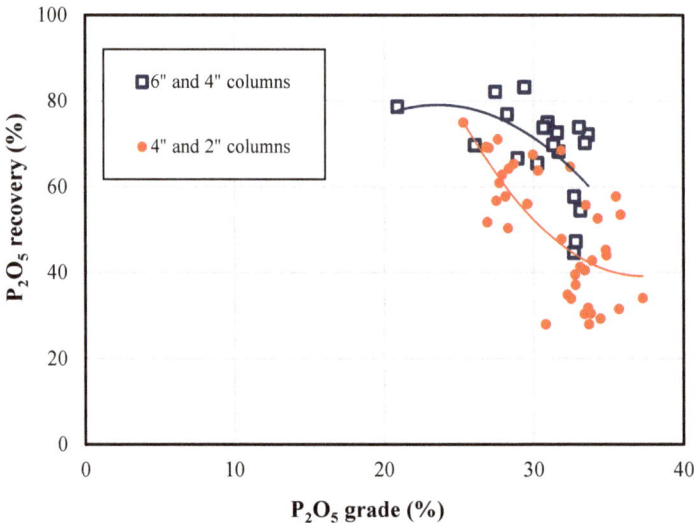

Figure 3. P2O5 grade/recovery curves for rougher/cleaner flotation. Comparative results of the circuit with 6" and 4" and 4" and 2" column flotation.

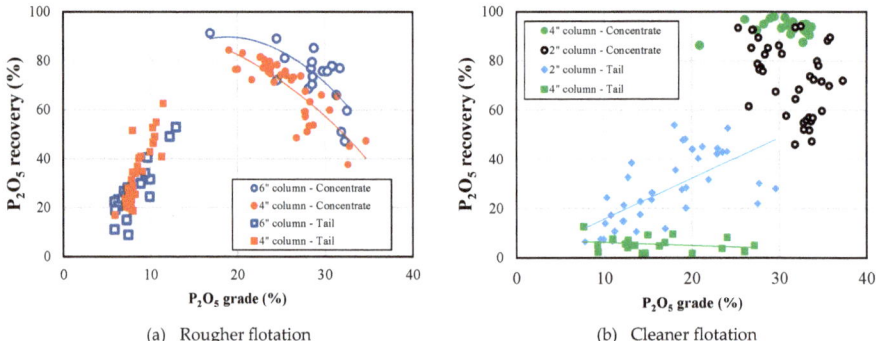

(a) Rougher flotation

(b) Cleaner flotation

Figure 4. P2O5 grade/recovery curves for rougher (**a**) and cleaner (**b**) stages. Comparative results for rougher flotation in 6" and 4" column flotation and for cleaner flotation in 4" and 2" column flotation.

As for cleaner flotation, it could be observed that the results obtained in the flotation with the 2" column showed a higher distribution compared to the results obtained for the 4" column. This can be explained by the fact that cleaner flotation in the 2" column had more process variables tested, focused on optimization of this stage, including changes in the superficial air velocity (Jg) and wash water (Jw), besides the reagents dosage variation. For the flotation in the 4" column as a cleaner, only the effect of the collector, depressant dosage, and pH were evaluated for Jg and the Jw kept fixed at 0.51 cm/s and 0.20 cm/s, respectively. The P_2O_5 losses on the cleaner flotation with the 4" column ranged from 2% to a maximum of 12%, considering the P_2O_5 grade variation from 9% to 25%. On the other hand, it could be observed that the cleaner flotation in the 2" column showed a much higher variation, especially for the P_2O_5 losses, which ranged from 10% to values up to 24% for the similar P_2O_5 content in the sink fraction.

The relation between the P_2O_5 grade in the final apatite concentrate and the content of Fe_2O_3 and SiO_2 impurities for both circuits evaluated are shown in Figure 5. The variation of the Fe_2O_3 content in the final apatite concentrate is very similar for both circuits and it is independent of the sample. The SiO_2 content has the same trend observed for the iron impurities in Sample 1 for both circuits, whereas the SiO_2 content in the final concentration of Sample 2 is significantly lower than that obtained for Sample 1. These results indicate that contamination of the apatite concentrate strongly depends on the ore characteristics that feed the plant and it is less influenced by the scale tested. As shown before, the SiO_2 grade in Sample 2 is much lower when compared with Sample 1, unlike the iron content, which is quite similar in both samples.

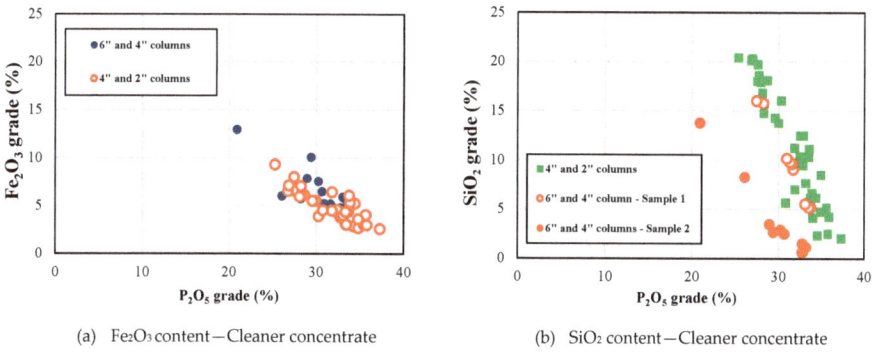

(a) Fe_2O_3 content—Cleaner concentrate (b) SiO_2 content—Cleaner concentrate

Figure 5. Relation between the P_2O_5 grade and Fe_2O_3 (a) and SiO_2 (b) impurities in the final apatite concentrate for both circuits. Flotation studies with Samples 1 and 2.

A summary of the best results achieved using rougher/cleaner configuration for the circuit with the 4" and 2" columns is shown in Table 4 for the flotation studies with Sample 1. Considering the average results for the three tests, mass recovery was around 23% and P_2O_5 recovery and grade were 54% and 35.2%, respectively. The SiO_2 and Fe_2O_3 impurities levels were 3.0% and 5.2%. Reagent consumptions were 117 g/t to the collector, around 2700 g/t to the depressant, and 410 g/t to NaOH and the pulp pH was fixed at 9.7.

Table 4. Summary of the best results obtained in experiments at 4" and 2" column flotation. Rougher/cleaner. Feed solids rate: 35 kg/h; pH = 9.7. Flotation studies with Sample 1.

Test	Recovery (%)		Grade (%)			Consumption (g/t)	
	Mass	P_2O_5	P_2O_5	Fe_2O_3	SiO_2	Collector	Depressant
1	24.2	52.7	34.3	2.9	6.2	123	2353
2	25.3	57.7	35.5	3.0	5.1	116	2737
3	22.7	53.5	35.8	3.0	4.2	113	3259
Average	24.1	54.6	35.2	3.0	5.2	117	2783

Table 5 shows the effect of pulp pH variation, for the values 9.7 and 10.8, on the flotation performance for the experiments using the 6" and 4" columns in rougher/cleaner configuration in experiments with Sample 1. The collector dosage was 130 g/t and the depressant was 2900 g/t. It can be observed that a reduction of the pulp pH from 10.8 to 9.7 decreases the P_2O_5 content in the concentrate, going from 33.4% to 31.4%. The reason for the dilution of the concentrate is exclusively caused by the increase of the SiO_2 from 5.3% to values around 10%, since the iron content is virtually the same. The increase in mass recovery, from 34.6% to 36.9% with the decrease of the pH value, is due the increase in the SiO$_2$ content once the P$_2$O$_5$ recovery is almost the same for both pH values evaluated, reaching 72%.

Table 5. Effect of pH variation on the flotation performance for experiments at 6" and 4" column flotation. Rougher/cleaner. Feed solids rate: 45 kg/h. Flotation studies with sample 1.

Test	Recovery (%)		Grade (%)			pH	Consumption (g/t)	
	Mass	P$_2$O$_5$	P$_2$O$_5$	Fe$_2$O$_3$	SiO$_2$		Collector	Depressant
1	35.0	72.1	33.7	5.2	5.2		132	3004
2	33.5	70.2	33.5	5.6	5.2	10.8	127	2905
3	35.3	73.8	33.1	5.9	5.5		129	2947
Average	34.6	72.0	33.4	5.6	5.3		129	2952
1	33.5	68.2	31.7	5.2	9.1		123	2815
2	37.9	72.6	31.6	4.9	9.7		125	2844
3	36.2	69.7	31.3	5.1	9.8	9.7	133	3029
4	39.9	74.9	31.0	5.2	10.2		130	2959
Average	36.9	71.3	31.4	5.1	9.7		127	2911

Flotation results for Sample 2 in the experiments carried, out with the 6" and 4" columns, are shown in Table 6. The average collector dosage for the four tests was around 80 g/t and the depressant was 2600 g/t at the pulp pH of 9.7. The average mass and P$_2$O$_5$ recovery were 27% and 51% for a P$_2$O$_5$ grade of 32.8%. Compared to the results obtained from Sample 1, the SiO$_2$ content in the final concentrate was much lower in the experiments with Sample 2, reaching an average grade of 1.0%. On the other hand, the iron content is very similar for both samples. The lower SiO$_2$ content reached in the concentrate of Sample 2 can be explained by the lower grade in the slime feed sample that was analyzed at 12.5%, compared with the 19.7% of Sample 1.

Table 6. Summary of the best results obtained in experiments at 6" and 4" column flotation. Rougher/cleaner. Feed solids rate: 45 kg/h. pH = 9.7. Flotation studies with Sample 2.

Test	Recovery (%)		Grade (%)			Consumption (g/t)	
	Mass	P$_2$O$_5$	P$_2$O$_5$	Fe$_2$O$_3$	SiO$_2$	Collector	Depressant
1	25.6	47.2	32.8	4.8	0.7	72	3000
2	24.6	44.6	32.7	3.8	0.6	69	2880
3	27.4	54.5	33.1	4.2	1.2	94	2380
4	30.7	57.7	32.7	4.4	1.5	87	2215
Average	27.1	51.0	32.8	4.3	1.0	80	2618

Regarding the reagent consumption (collector and depressant), it was observed that the average collector dosage is similar to the levels applied at the industrial scale in Brazilian plants. On the other hand, the depressant dosage is much higher, reaching values up to 2500 g/t. Usually, the depressant dosage does not exceed 1200–1300 g/t. Matiolo et al [15] evaluated the depressant dosage on apatite flotation from slimes. The results indicated that both parameters (P$_2$O$_5$ grade and recovery) improve when the depressant dosage increases from 1230 g/t to values reaching higher than 1700 g/t. It was also found that to control the iron impurities in the final flotation concentrate, the depressant dosage must be up 2200 g/t to a maximum of 3000 g/t.

A similar approach to that evaluated in this study of recovering the valuable fine phosphate particles (<45 μm) from their slimes through the application of the column flotation technique was tested by Abdel-Khalek [20]. Tests were performed using oleic acid as a collector for the phosphate minerals and sodium silicate as a depressant for their associated gangues. The main operating parameters affecting the performance of column flotation were investigated. The results indicate that the best operating conditions for column flotation of phosphate slimes are as follows: A superficial gas velocity of 0.84 cm/s, a frother concentration of 0.1 kg/ton, a column height of 230.5 cm, and a superficial water velocity of 2.2 cm/s. Under these conditions, a product assaying 25.3% P_2O_5 and 14.64% I.R. (insoluble residue), with a P_2O_5 recovery of about 51.52%, is obtained from a feed containing 18.26% P_2O_5 and 24.03% I.R. Such grades and recoveries are not obtained by applying the conventional froth flotation technique, even after cleaning the rougher concentrate. Recovery of more than 50% of phosphate from disposed slimes will improve the economic viability of the beneficiation process for phosphate ores. It will also help to solve the environmental problems associated with the disposal of these slimes.

5. Conclusions

It was possible to obtain apatite concentrates from two different slime samples with P_2O_5 grades higher than 33%, which can be applied to SSP (single superphosphate) and TSP (triple superphosphate) production. Considering desliming and flotation, the overall mass recovery ranged from 9% to 12% and P_2O_5 recovery ranged between 29–42%. It is estimated that the earnings obtained by recovery of the apatite from slimes may represent an increase in 4% of the overall P_2O_5 recovery at industrial plants. The results obtained in this study indicate that the flotation results were better in the circuits with higher column flotation (6" and 4"). The results can be explained by the application of a cavitation tube in the rougher stage in the 6" column. The improved flotation performance can be attributed to increased probabilities of collision and attachment and the reduced probability of detachment by the small size bubbles generated by the cavitation tube in comparison to the bubbles produced by the porous tube of the 4" column flotation.

Author Contributions: Conceptualization, E.M., H.J.B.C and M.F.d.L.T; methodology, E.M., H.J.B.C and M.F.d.L.T.; formal analysis, E.M., H.J.B.C., M.F.d.L.T., R.N.d.A and A.S.d.F; investigation, E.M., H.J.B.C., M.F.d.L.T., R.N.d.A and A.S.d.F; resources, E.M., H.J.B.C. and M.F.d.L.T; data curation, E.M., H.J.B.C., M.F.d.L.T., R.N.d.A and A.S.d.F; writing—original draft preparation, E.M., H.J.B.C. and A.S.d.F; writing—review and editing, E.M., H.J.B.C., M.F.d.L.T., R.N.d.A and A.S.d.F; visualization, E.M., H.J.B.C., M.F.d.L.T., R.N.d.A and A.S.d.F; supervision, E.M., H.J.B.C., and M.F.d.L.T.; project administration, E.M., and M.F.d.L.T.,; funding acquisition, E.M., and M.F.d.L.T.

Funding: This research received no external funding.

Acknowledgments: The authors would like to thank the Brazilian Institutes supporting this research, namely CNPq for the scholarship to Amanda and Renata, and CETEM/MCTIC for the infrastructure. A special thanks to Copebras (CMOC International) for the financial support and permission to publish this work, and to process technician, Fabio Novaes, for his support during the experimental work.

Conflicts of Interest: The authors declare no conflict of interest.

References

1. Houot, R. Beneficiation of phosphatic ores through flotation: Review of industrial applications and potential developments. *Int. J. Miner. Process.* **1982**, *9*, 353–384. [CrossRef]
2. Dong, X.; Liu, S.; Yao, Y.; Liu, H.; Pei, Y. A review of new technological progress for benefication of refractory phosphate ore in China. *IOP Conf. Ser.: Earth Environ. Sci.* **2017**, *63*, 1–6.
3. Pourkarimi, Z.; Rezai, B.; Noaparast, M. Nanobubbles effect on the mechanical flotation of phosphate ore fine particles. *Physicochem. Probl. Miner. Process.* **2018**, *54*, 278–292.
4. Huynh, L.; Feiler, A.; Michelmore, A.; Ralston, J.; Jenkins, P. Control of slime coatings by the use of anionic phosphates: A fundamental study. *Min. Eng.* **2000**, *13*, 1050–1069. [CrossRef]
5. Ahmed, H.A.M. Optimization of desliming prior to phosphate ore upgrading flotation. *Phys. Probl. Min. Proc.* **2007**, *41*, 79–88.
6. Teague, A.J.; Lollback, M.C. The beneficiation of ultrafine phosphate. *Min. Eng.* **2012**, *27–28*, 52–59. [CrossRef]
7. Al-Thyabat, S. Evaluation of mechanical flotation of non-slimed Jordanian siliceous phosphate. *Arab. J. Sci. Eng.* **2012**, *31*, 877–887. [CrossRef]
8. Zhang, P.; Bogan, M. Recovery of phosphate from Florida beneficiation slimes I. Re-identifying the problem. *Min. Eng.* **1995**, *8*, 523–534. [CrossRef]

9. Pradip, R.S.; Sankar, T.A.P. Selective flotation of Maton (India) phosphate ore slimes with particular reference to the effects of particle size. *Internet J. Min. Proc.* **1992**, *36*, 283–293.
10. Al-Thyabat, S.; Yoon, R.H.; Shin, D. Floatability of fine phosphate in a batch column flotation cell. *Min. Metal. Proc.* **2011**, *28*, 1110–1116. [CrossRef]
11. Liu, X.; Zhang, Y.; Liu, T.; Cai, Z.; Sun, K. Characterization and separation studies of fine sedimentary phosphate ore slime. *Minerals* **2017**, *7*, 94. [CrossRef]
12. Guimarães, R.C.; Peres, A.E.C. Industrial practice of phosphate ore flotation at Serrana-Araxá, Brazil. In Proceedings of the XXI International Mineral Processing Congress, Rome, Italy, 23–27 July 2000; B9-17.
13. Guimarães, R.C.; Peres, A.E.C. Experiência brasileira de produção de concentrado fosfático a partir de lamas. In Proceedings of the XIX Encontro Nacional de Tratamento de Minérios e Metalurgia Extrativa, Recife, Brasil, 26–29 November 2002; Volume I, pp. 247–253. (In Portuguese).
14. Guimarães, R.C.; Peres, A.E.C. Production of phosphate concentrates from slimes: Brazilian experience. In Proceedings of the XXII International Mineral Processing Congress, Cape-Town, South Africa, 29 September–3 October 2003; pp. 606–612.
15. Matiolo, E.; Couto, H.J.B.; Teixeira, M.F.L.; Freitas, A.S.; Almeida, R.N. Recovery of apatite from slimes of a Brazilian phosphate ore. *J. Wuhan Inst. Technol.* **2017**, *39*, 39–48.
16. Ipek, H.; Ozdag, H. An investigation into the enrichment of phosphate slime by column flotation. *Dev. Miner. Process.* **2000**, *13*, C8a-1–C8a-5.
17. Tao, Y.; Liu, J.; Yu, S.; Tao, D. Picobubble enhanced fine coal. *Sep. Sci. Technol.* **2006**, *41*, 3597–3607. [CrossRef]
18. Zhou, Z.A.; Xu, Z.; Fich, J.A.; Hu, H.; Rao, S.R. Role of hydrodynamic cavitation in fine particle flotation. *Int. J. Miner. Process.* **1997**, *51*, 139–149. [CrossRef]
19. Zhou, Z.A.; Xu, Z.; Finch, J.A.; Masliyah, J.H.; Chow, R.S. On the role of cavitation in particle collection in flotation—A critical review. II. *Min. Eng.* **2009**, *22*, 419–433. [CrossRef]
20. Abdel-Khalek, N.A.; Hassan, F.; Arafa, M.A. Separation of valuable fine phosphate particles from their slimes by column flotation. *Sep. Sci. Tech.* **2000**, *35*, 1077–1086. [CrossRef]

© 2019 by the authors. Licensee MDPI, Basel, Switzerland. This article is an open access article distributed under the terms and conditions of the Creative Commons Attribution (CC BY) license (http://creativecommons.org/licenses/by/4.0/).

Article

Rare Earth Occurrences in Streams of Processing a Phosphate Ore

Xiaosheng Yang [1,*], Hannu Tapani Makkonen [1] and Lassi Pakkanen [2]

1 Geological Survey of Finland, Outokumpu 83500, Finland; hannu.t.makkonen@gtk.fi
2 Geological Survey of Finland, Espoo 02151, Finland; lassi.pakkanen@gtk.fi
* Correspondence: jason.yang@gtk.fi; Tel.: +358-50348-6063

Received: 11 February 2019; Accepted: 27 April 2019; Published: 30 April 2019

Abstract: Rare earth elements (REEs) are defined as lanthanides with Y and Sc. Rare earth occurrences including the REE-bearing phases and their distributions, measured by rare earth oxides (REOs), in the streams of processing a phosphate ore were determined by using MLA, the mineral liberation analysis and EPMA, the electron probe microanalysis. The process includes an apatite ore beneficiation by flotation and further processing of the beneficiation concentrate with sulfuric acid. Twenty-six, sixty-two and twelve percent of the total REOs (TREO) contents from the ore end up in the products of beneficiation tailings, phosphogypsum (PG) and phosphoric acid, respectively. Apatite, allanite, monazite and pyrochlore are identified as REE-bearing minerals in the beneficiation process. In the beneficiation tailings, the REEs are mainly distributed in monazite (10.3% TREO), apatite (5.9% TREO), allanite (5.4% TREO) and pyrochlore (4.3% TREO). Gypsum, monazite, apatite and other REE-bearing phases were found to host REEs in the PG and the REEs distributions are 44.9% TREO in gypsum, 15.8% TREO in monazite, 0.6% TREO in apatite and 0.6% TREO in other REE-bearing phases. Perspectives on the efficient recovery of REEs from the beneficiation tailings and the PG are discussed.

Keywords: rare earth elements (REEs); phosphate ore processing; REE-bearing phases; beneficiation tailings; phosphogypsum (PG)

---

1. Introduction

The grade of apatite ores can vary from 4% to 20% phosphorus pentoxide (P_2O_5). Normally, beneficiation by flotation is needed to obtain a high grade (often more than 30% P_2O_5) apatite concentrate. Meanwhile, beneficiation tailings including waste clay are produced. Then phosphoric acid is prepared in the wet-process by adding sulfuric acid to the apatite concentrate. Phosphogypsum (PG) is the major solid product formed during the process. Rare earth elements (REEs) comprising 15 lanthanides plus Y and Sc are normally contained in the phosphate rocks in various contents from 0.03 to 1.0 wt. % [1,2]. In the phosphate rock processing, most of the REEs from the phosphate rock (over 85%) end up in the wastes (waste clay, flotation tailings and phosphogypsum) [3]. Because these wastes have very fine grain sizes and complex mineralogical compositions, the reprocessing of them for economic recovery of REEs becomes very challenging [3]. Research was carried out to verify whether the wastes might be useful as a raw material for REEs recovery [4–12] and the manufacture of building materials [1]. However, the extraction of REEs and other valuable elements from these wastes in phosphate processing has not been industrially realized, while trace elements are considered one of the main environmental concerns of direct use [13].

The incorporation of REEs in PG from solution during the wet-process of phosphoric acid production using sulfuric acid has been investigated for phosphate rocks with different REE grades and it is concluded that about 60–70% of the initial REEs content was lost into the PG [14–20]. Meanwhile, the occurrence modes or the precipitated phases of REEs in the PG are considered but conclusions are still disputable. Jin et al. (2011) [21] concluded that REEs were present as tiny crystals of rare earth sulphates. A study by Santos et al. (2006) [22] showed that although REEs were enriched in the PG samples they were not associated with CaSO4 itself. The study by Borges et al. (2016) [23] indicated that REEs occur as sulphates, carbonates, fluorides and phosphates. Schmidt et al. (2009) [24] concluded that Eu^{3+} was adsorbed as an inner-sphere species on the precipitated gypsum. In addition, REEs including Y, La, Ce, Nd, Sm and Pr were detected on the PG surface in both elemental as well as oxide forms by TOF-SIMS [6]. Alhassanieh et al. (2012) [25], in contrast, concluded that Eu^{3+} was incorporated in the structure of the gypsum as opposed to being adsorbed on the gypsum surface. The study by Dutrizac, J.E. (2017) [14] on the behaviour of all the trivalent REEs concluded that the rare earth ions are structurally incorporated in gypsum, according to a mechanism involving the replacement of three Ca^{2+} ions by two REE^{3+} ions and a vacant Ca^{2+} lattice site.

Compared to the studies on the incorporation and the occurrence modes of REEs in the PG during the wet-process of phosphoric acid production, the studies about REEs transferring and their occurrences in the beneficiation process are very limited. A study by Zhang et al. (2018) [3] indicated that REEs during a phosphate ore processing are distributed approximately 40% in waste clay, 10% in amine tails, 38% in PG and 12% in phosphoric acid but how the REEs occur in these products was not investigated.

In the present study, the occurrences of REEs, including REE-bearing phases and REEs distribution, in the streams of processing a phosphate rock were studied on the basis of mineralogical analyses using MLA, the mineral liberation analysis and EPMA, the electron probe microanalysis of the apatite ore, the flotation concentrate and the PG. Potential technological pathways for REE recovery from the beneficiation tailings and the PG are discussed.

2. Materials and Methods

The samples of apatite ore, flotation concentrate, and PG were obtained from a phosphate mine, which is a carbonatite deposit with an in-situ grade of around 5.0% P_2O_5. The general concentration process is that the ore is crushed and ground first and then flotation is performed to obtain the apatite concentrate. The beneficiation concentrate is then treated using sulfuric acid for phosphoric acid production. At the same time the PG is produced as the leaching residue. Currently, the material of PG in the mine is not used commercially. Because the beneficiation is a physical process, the REE-bearing phases in the concentrate and the tailings are the same as in the ore.

The samples were split into 100 g subsamples for chemical and mineralogical analyses. The PG sample was classified by screen into three fractions (+75 µm, −75 + 45 µm and −45 µm) and was analysed separately.

Leaching experiments of the flotation concentrate with sulfuric acid were carried out at the conditions of sulfuric acid concentration 2.5 M, temperature 60 °C, acid solution 500 mL, solid 150 g and agitation speed 400 rpm.

Sodium peroxide plus sodium hydroxide digestion was used for the determination of REEs using the ICP-MS, the inductively coupled plasma mass spectrometry, technique, which was performed by Labtium Oy in Finland. The limits of detection (LODs) of the ICP-MS for REEs are shown in Table 1.

Table 1. The limits of detection (LODs) of the ICP-MS for REEs (mg/kg).

La	Ce	Pr	Nd	Sm	Eu	Gd	Tb	Dy	Ho	Er	Tm	Yb	Lu	Y	Sc
0.1	0.1	0.1	0.2	0.2	0.05	0.05	0.1	0.1	0.1	0.05	0.1	0.05	0.1	1	1

The modal mineralogy (i.e., the percentages of the mineral components), the mineral liberation and the grain size distribution of REE-bearing minerals were measured by using a mineral liberation analyser (MLA) (a FEI MLA Quanta 600 system) at the Geological Survey of Finland in Outokumpu. Minerals were identified mostly based on EDS analyses. Using the EDS analysis, the composition of a certain mineral was determined, and the composition was compared to the composition of the mineral as specified in the MinIdent mineralogy database and a mineralogical book [26]. Also, the EDS spectrum was compared to the spectra in the databases of MLA.

For determining the chemical compositions of REE-bearing phases in the samples, and also of certain minerals that cannot be unequivocally identified by the semi-quantitative approach of MLA, the electron probe microanalysis (EPMA) was performed by the wavelength dispersive technique using a Cameca SX100 instrument at the Geological Survey of Finland in Espoo. All analyses were determined using an accelerating voltage of 20 kV. The probe currents and beam diameters used were 6–60 nA and 1–10 µm, respectively. Analytical results were corrected using the PAP on-line correction programme [27]. Depending on samples and minerals the limits of detection (LODs) of the EPMA for REEs are shown in Tables 2 and 3.

Table 2. The limits of detection (LODs) of the EPMA for REEs with sample phosphogypsum (PG) (mg/kg).

REE-Bearing Phase	La	Ce	Pr	Nd	Sm	Gd	Tb	Dy	Ho	Er	Yb	Lu	Y
Gypsum	184	170	247	340	223	246							70
Apatite	176	148	262	340	230	233							79
Monazite	481	329	545	640	253	527	362	378	532	551	290	222	426
Other phase	185	183	259	362	251	275							85

Table 3. The limits of detection (LODs) of the EPMA for REEs with samples of ore and apatite concentrate (mg/kg).

REE-Bearing Phase	La	Ce	Pr	Nd	Sm	Gd	Dy	Y	Sc
Apatite	983	1100	1576	2194	2138	2027	1077	930	
Allanite	1353	1534	2022	2678	2700	2460	1411	1126	
Monazite	1353	1534	2022	2678	2700	2460	1411	1126	
Zircon	n.d. *	1302	1846	2569	2515	2368	1243	1060	421
Pyrochlore	1721	1821	2571	3644	3458	3247	1794	1640	603

* not determined.

It should be stressed that, because the mineral identification was carried out using the semi-quantitative EDS method, phases with complex chemical compositions, such as allanite, aeschynite, britholite and pyrochlore, should in fact be referred to as phases with allanite-like, aeschynite-like, and so forth, compositions. However, although identification of such phases cannot be treated as entirely accurate, for the sake of clarity and readability of the text simplified and more general names will be used.

The analytical steps and strategy are presented in Figure 1. By combining MLA and EPMA analyses, the REEs concentrations contributed by all REE-bearing phases in the samples are determined. Then, the REE occurrences, for example, the weight distributions of REEs in the streams of the phosphate rock processing with REE-bearing phases, are achieved by combining experimental and commercial production data.

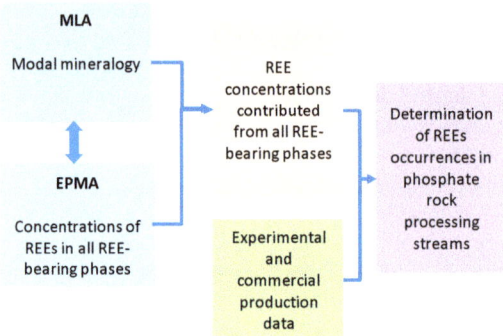

Figure 1. The analytical steps and strategy.

3. Results

3.1. The Individual Concentrations of REEs

The individual concentrations of REEs (measured by rare earth oxides (REOs)) of the ore, the flotation concentrate and tailings and the PG analysed by ICP-MS are shown in Table 4. All the REEs except Pm were detected in the samples but eight REEs, La, Ce, Pr, Nd, Sm, Gd, Dy and Y, are found with relatively high concentrations. Other REEs are in trace contents (REOs < 5 mg/kg). The contents that are smaller than the limits of detection are marked by n.d. (not detected). The contents of total REOs (TREO) are: 539 mg/kg in the ore; 3431 mg/kg in the concentrate; 255 mg/kg in the tailings; and 2280 mg/kg in the PG. It is noted that for the PG the REEs are not evenly distributed in the size fractions and the fine fraction (−45 μm) has higher contents of TREO as shown in Table 5.

Table 4. The individual REEs concentrations of the ore, concentrate and PG by ICP-MS (mg/kg).

	La_2O_3	Ce_2O_3	Pr_2O_3	Nd_2O_3	Sm_2O_3	Eu_2O_3	Gd_2O_3	Tb_2O_3	Dy_2O_3
Ore	89.60	235.43	30.08	119.32	17.80	4.48	12.39	1.27	5.11
Flot Conc	581.81	1488.11	201.08	800.22	116.46	28.44	78.52	7.69	29.02
Flot Tail *	41.23	112.33	13.27	52.41	8.11	2.13	5.89	0.64	2.76
PG	339.75	1029.09	134.00	547.27	81.87	19.57	51.87	4.95	17.44
	Ho_2O_3	Er_2O_3	Tm_2O_3	Yb_2O_3	Lu_2O_3	Sc_2O_3	Y_2O_3	TREO	
Ore	0.76	1.79	0.18	1.00	n.d.	10.00	19.68	539.02	
Flot Conc	4.09	8.95	0.82	4..11	0.50	9.50	81.28	3431.11	
Flot Tail *	0.43	1.09	n.d.	0.69	n.d.	n.d.	13.63	254.83	
PG	2.41	4.92	0.34	1.15	n.d.	n.d.	45.72	2280.56	

* Calculated based on chemical analysis results of ore and concentrate (the concentrate mass yield is 8.95 wt. % according to the concentrator production record).

Table 5. The individual REEs concentrations of in the size fractions of +75, −75 + 45 and −45 μm in the PG (mg/kg).

	La_2O_3	Ce_2O_3	Pr_2O_3	Nd_2O_3	Sm_2O_3	Eu_2O_3	Gd_2O_3	Tb_2O_3	Dy_2O_3
+75 μm	290.85	930.0	126.39	528.37	79.55	18.99	50.14	4.83	17.33
−75 + 45 μm	261.53	839.81	113.29	473.55	73.06	17.48	45.53	4.49	15.61
−45 μm	485.53	1358.69	166.19	652.01	94.62	22.70	60.97	5.76	19.63
	Ho_2O_3	Er_2O_3	Tm_2O_3	Yb_2O_3	Lu_2O_3	Sc_2O_3	Y_2O_3	TREO	
+75 μm	1.83	4.80	0.23	1.25	n.d.	n.d.	41.65	2096.34	
−75 + 45 μm	2.29	4.35	0.34	1.25	n.d.	n.d.	41.15	1893.84	
−45 μm	2.06	3.89	0.23	0.91	n.d.	n.d.	36.45	2909.74	

The weight percentages of three size fractions of the PG are: +75 μm 36.1 wt. %, −75 + 45 μm 33.7 wt. % and −45 μm 30.2 wt. %.

3.2. REE-bearing Phases and Distributions of REEs

3.2.1. Beneficiation Process

The modal mineralogy of the apatite ore and concentrate by MLA are shown in Table 6. Apatite, allanite, monazite, zircon and pyrochlore were found to be REE-bearing minerals which contain measurable amounts of REEs. Apatite was enriched from a content of 8.9%wt in the phosphate rock to that of 88.5%wt in the concentrate, but other REE-bearing minerals were not enriched during flotation. Because flotation is a physical process mineral phases are not changed. Thus, the modal mineralogy of the tailings can be calculated based on analysed data of the ore and concentrate.

Table 6. Modal mineralogy of the ore and flotation concentrate by mineral liberation analyser (MLA).

	Mineral	Phosphate Rock, wt. %	Apatite Concentrate, wt. %	Tailings, wt. % *
REE-bearing minerals	Apatite	8.880	88.460	1.052
	Monazite	0.003	0.004	0.003
	Allanite	0.0051	0.0046	0.005
	Zircon	0.018	0.008	0.019
	Pyrochlore	0.003	0.001	0.003
non REE-bearing minerals	Calcite	16.680	8.700	17.460
	Ankerite	0.081	0.140	0.075
	Dolomite	2.450	1.350	2.561
	Other silicates	70.780	1.020	77.64

* calculated data based on analysed data of ore and concentrate. The mass yields of the concentrate and tailings are 8.95% and 91.05% according to the concentrator production record.

These REE-bearing minerals identified by MLA were taken into account in the EPMA analyses from which the concentrations of REEs in all the REE-bearing minerals (measured by REOs) were determined (Table 7). The data in the table are the average calculated data of multiple analyses. The numbers of analyses are stated in Table 7.

Eight REEs, including La, Ce, Pr, Nd, Sm, Gd, Dy and Y, were measured for all five REE-bearing minerals. The REOs marking as n.d. in the table are below the LODs. It noted that Sc was detected by ICP-MS in the ore and flotation concentrate shown in Table 4. According to the authors' previous experience and a study [28] zircon and pyrochlore might be the Sc-bearing phases. Thus, Sc was measured by EPMA for zircon and pyrochlore but was not detected (below the LOD) because the EPMA has higher LOD than ICP-MS. For apatite, allanite and monazite Sc was not measured.

Table 7. The individual concentrations of REEs in all the REE-bearing phases: apatite, allanite, monazite, zircon and pyrochlore by EPMA (mg/kg) (measured by REOs).

	Apatite	Allanite	Monazite	Zircon	Pyrochlore
Number of analyses	20	9	11	9	7
La_2O_3	n.d.	51724	173018	n.d.	35183
Ce_2O_3	2133	127800	381318	n.d.	149817
Pr_2O_3	n.d.	12401	32844	n.d.	19167
Nd_2O_3	n.d.	35174	95177	n.d.	72237
Sm_2O_3	n.d.	n.d.	5425	n.d.	10919
Gd_2O_3	n.d.	n.d.	n.d.	n.d.	5010
Dy_2O_3	n.d.	n.d.	n.d.	n.d.	n.d.
Sc_2O_3				n.d.	n.d.
Y_2O_3	n.d.	n.d.	n.d.	n.d.	1459
TREO	2133	227099	687782	0	293792

The concentrations of REOs contributed by each REE-bearing mineral in the ore, the concentrate and the tailings are obtained from the analysis results of MLA and EPMA and presented in Figures 2–4, respectively.

For the ore with over 81% of REEs are in apatite and other REE-bearing minerals: monazite, allanite and pyrochlore carry about 19% of REEs. Zircon was identified the REE-bearing mineral but the contents of all REOs were below the LODs. The REEs in apatite are mainly Ce and other REEs are below the LODs. The REEs, such as La, Nd, Pr and Y, are present in significant amounts in monazite, allanite and pyrochlore.

During flotation, apatite was effectively enriched in the concentrate. For the concentrate, almost all (98%) of REEs are in apatite. In contrast, for the tailings nearly 80% of REEs are carried by monazite, allanite and pyrochlore. That is, REEs are lost into the tailings mainly due to the fact that monazite, allanite and pyrochlore have poorer flotation efficiency compared to apatite.

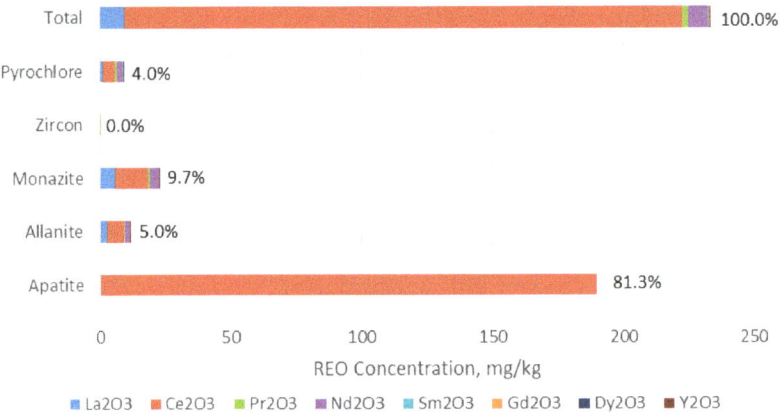

Figure 2. Rare Earth oxide (REO) concentrations in the REE-bearing phases of the ore (mg/kg).

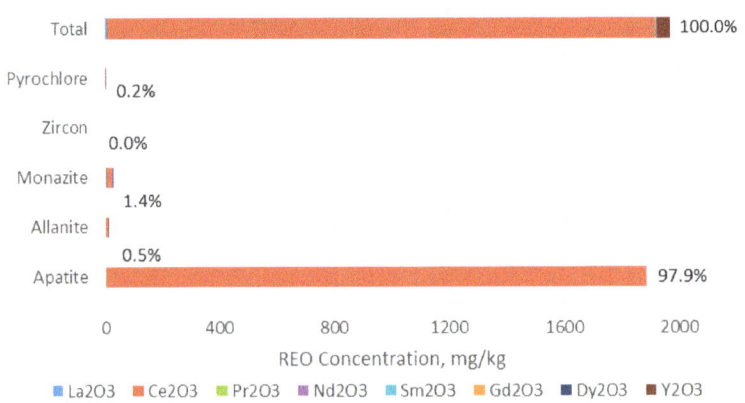

Figure 3. REO concentrations in the REE-bearing phases of the concentrate (mg/kg).

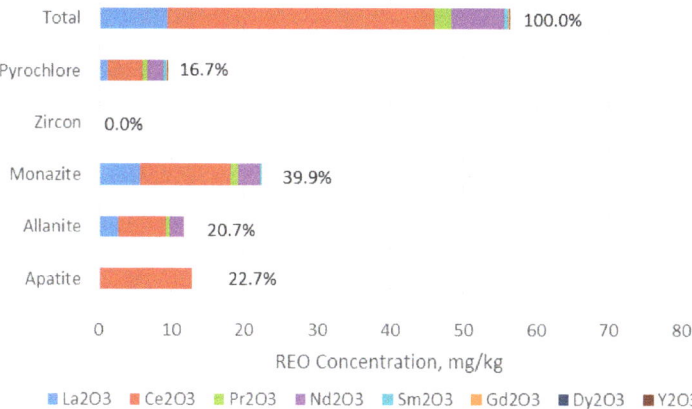

Figure 4. REO concentrations in the REE-bearing phases of the tailings (mg/kg).

According to the concentrator production record in the flotation process, the mass yields of concentrate and tailings are 8.95% and 91.05%, respectively. The recoveries of REEs measured by TREO content in the concentrate and the tailings from the ore for different REE-bearing phases were calculated based on the data of REO concentrations shown in Figures 2–4 and in Table 8. Over 89% of TREO were recovered in the concentrate from apatite but much less from allanite, monazite and pyrochlore, that is, most of REEs from allanite, monazite and pyrochlore are lost into the tailings. The overall recoveries of TREO content in the concentrate and the tailings are 74.1% and 22.0%, respectively. It is noted that because of different detection limits and accuracies of MLA and EPMA the summations are not exactly 100% but in the range of 95.0% to 100.8%, which are at very high precisions.

Table 8. Recoveries of total rare earth oxide (TREO) in the concentrate and tailings from different REE-bearing minerals.

	Apatite	Allanite	Monazite	Pyrochlore	Total Rec
Concentrate	89.2	8.2	10.8	2.8	74.1
Tailings	6.1	91.9	90.0	92.2	22.0
Sum	95.3	100.0	100.8	95.0	96.1

3.2.2. The Phosphogypsum (PG)

Modal mineralogy of the PG sample in three size fractions calculated by MLA are shown in Table 9. REE-bearing minerals including monazite, aeschynite and britholite were identified. Monazite (0.01–0.12%) is the prevailing REE mineral. The other two occur only in trace amounts. Monazite is more concentrated in the fine fraction of −45 μm and the content reaches 0.12 wt. %. The concentrations of monazite in the fractions of +75 μm and −75 + 45 μm are almost the same, only 0.01–0.02 wt. %. The distributions of monazite in the size fractions is consistent with the REE distribution in these fractions as shown in Table 5.

The mineralogical analysis indicates that the grain size of monazite is around 20 μm and the liberation degrees of monazite (>95% liberated) in the fractions of −45 μm, −75 + 45 μm and +75 μm are 100%, 95% and 85%, respectively.

Gypsum ($Ca(SO_4)\cdot 2(H_2O)$) is, naturally, the dominating mineral with the average content of 97.5 wt. % which was detected by EPMA to be a REE-bearing phase. The other REE-bearing phase listed in the table is not a clear mineral identified by MLA and EPMA.

Table 9. Modal mineralogy of the PG sample (wt. %).

	Mineral	+75 μm	−75 + 45 μm	−45 μm
REE-bearing phases	Gypsum	96.513	99.017	97.094
	Apatite	0.802	0.055	0.119
	Monazite	0.011	0.017	0.119
	Other REE-bearing phase	0.061	0.031	0.211
Non REE-bearing phases	Celestine	0.123	0.0545	0.475
	Fluorite_mix	1.249	0.212	0.867
	Tremolite	0.334	0.136	0.149
	Other (silicates, Fe-oxide, etc.)	1.121	0.493	0.938
	Total	100	100	100

The individual concentrations of REEs (measured by REOs) in all the REE-bearing phases by EPMA are shown in Table 10 which are the averages of many measurements. The REOs marking as n.d. in the table are below the LODs. Based on these data in Table 10 and the data of modal mineralogy in Table 9 the REO concentrations associated and distributions with these phases are calculated and shown in Figure 5. It is revealed that the TREO concentration in the PG is 0.16%wt and almost all (>98%) of REEs are shared by gypsum and monazite. Gypsum is the largest carrier of REEs which holds over 73% of REO content and 25% of REO content is contained in monazite. The REO contents shared by apatite and the other REE-bearing phase are less than 2%.

Table 10. The individual concentrations of REEs (measured by REOs) of the REE-bearing phases (mg/kg) *.

	Gypsum	Apatite	Monazite	Other REE-Bearing Phase
Number of analyses	89	37	8	5
La_2O_3	n.d.	506	192,697	2176
Ce_2O_3	592	1587	393,139	6358
Pr_2O_3	n.d.	n.d.	31,134	776
Nd_2O_3	342	1025	77,262	2444
Sm_2O_3	n.d.	n.d.	4124	453
Gd_2O_3	n.d.	n.d.	856	280
Y_2O_3	n.d.	179	n.d.	235
TREO	934	3788	699,212	122,622

* Certain elements in empty fields were not detected.

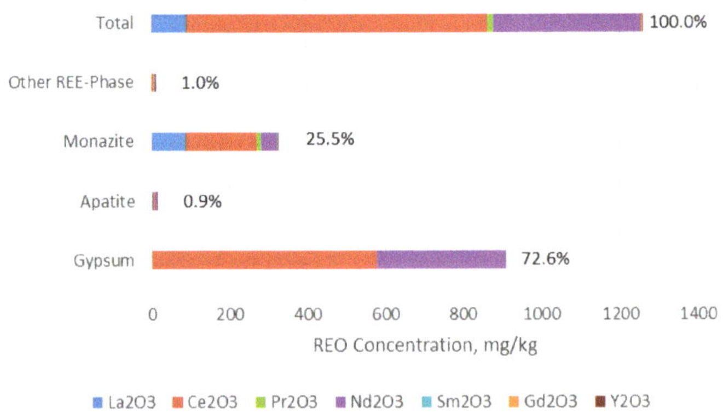

Figure 5. REO concentrations in REE-bearing phases (mg/kg) and recoveries (%).

A leaching experiment of the apatite concentrate with sulfuric acid under conditions of sulfuric acid concentration of 2.5 M and a temperature 60°C showed the yield of PG by weight (ratio of PG weight to feed weight) in the leaching process is 128%.

The precipitation rates of REEs from the concentrate into the PG were calculated by combining the data of REO concentrations shown in Figure 5 and the yield of PG in the leaching process. The results are shown in Table 11. It is indicated that, depending on the element, the calculated precipitation rates from the concentrate into the PG are in the range of 28.2% to 90.0% REO. The overall precipitation of TREO is 85.1%. Because the recovery of TREO in the concentrate is 74.1% the recovery of TREO in the PG from the ore is 63.0%.

Table 11. The precipitation rates of REO from the concentrate into the PG (%).

La_2O_3	Ce_2O_3	Pr_2O_3	Nd_2O_3	Sm_2O_3	Eu_2O_3	Gd_2O_3	Tb_2O_3
74.75	88.52	85.30	87.54	89.98	88.08	84.56	82.39
Dy_2O_3	Ho_2O_3	Er_2O_3	Tm_2O_3	Yb_2O_3	Lu_2O_3	Y_2O_3	TREO
76.92	75.42	70.36	53.07	35.82	28.16	72.00	85.08

3.3. Occurrences of REEs in the Processing Streams

From the analyses shown, assuming the phosphate mine processes 1.0 Mt of the phosphate rock annually, the occurrences and distributions of REEs (measured by TREO contents) among the processing streams are summarized in Table 12. It is shown that during the processing, 26% and 62% of the TREO contents from the rock, respectively, end up in the beneficiation tailings and the PG. In the tailings the REEs mainly occur in the phases of apatite, monazite, allanite and pyrochlore and in the PG the REEs largely occur in the phases of gypsum and monazite. Only 12% of them are recovered from the product of phosphoric acid in the form of ions.

Table 12. The Occurrences of REO in the Processing Streams.

Product	Weight, t	TREO, mg/kg	REEs Occurrence		
			REE-Bearing Phase	TREO, t	%
Phosphate Rock	1,000,000	233		233	100.0
			Apatite	189.3	81.3
			Allanite	11.5	5.0
			Monazite	22.7	9.7
			Pyrochlore	9.3	4.0
Concentrate	89,500	1928		172.5	74.1
			Apatite	168.8	72.5
			Allanite	0.9	0.4
			Monazite	2.5	1.1
			Pyrochlore	0.3	0.1
Tailings	910,500	66		60.3	25.9
			Apatite	13.7	5.9
			Allanite	12.5	5.4
			Monazite	24.0	10.3
			Pyrochlore	10.1	4.3
Phosphogypsum (PG)	114,560	1258		144.1	61.9
			Gypsum	104.6	44.9
			Apatite	1.3	0,6
			Monazite	36.7	15.8
			Other REE-bearing phase	1.4	0.6
Phosphoric acid	?	?	Ions in solution	28.4	12.2

4. Discussion

In this study both MLA and EPMA analyses were used to determine REEs occurrence in the products and the wastes generated during the processing of apatite ore. Compared to using only MLA the combination of MLA and EPMA analyses provides more precise analysis data because of trace contents of REEs and complexity REE-bearing phases in these products and wastes. For instance, for this studied case MLA only identifies the occurrences of less than 20% of REOs in the PG. Over 80% of REOs existing in the phase of gypsum cannot be recognized by this method. EPMA is able to acquire more precise and quantitative elemental analyses of all REE-bearing phases. Thus, the distributions of REEs in the products of ore beneficiation process and concentrate wet-process can be determined and the values of these products on REEs recovery can be quantitatively assessed.

Compared to ICP-MS analysis the MLA-EPMA method can quantify the chemical compositions of mineralogical phases by measuring solid materials at the micrometre scale. The measured data of Ce_2O_3 and TREO for four different samples by ICP-MS and MLA-EPMA methods are compared in Table 13.

As shown in Table 13, the data of Ce_2O_3 and TREO measured with two methods are comparable. For the trace REEs in these samples there are large deviations between the measured data of their oxides by two methods because of significant differences on the detection limits of elements.

Table 13. The comparison of the measured data of Ce_2O_3 and TREO for four different samples by ICP-MS and MLA-EPMA methods (mg/kg).

	Ore		Concentrate		Tailings		PG	
	Ce_2O_3	TREO	Ce_2O_3	TREO	Ce_2O_3	TREO	Ce_2O_3	TREO
MLA-EPMA	213	233	1909	1928	37	56	770	1258
ICP-MS	235	539	1488	3431	112	255	1029	2281

The outcomes presented in Table 12 provide not only the basic data to judge the values of these products and wastes generated during the apatite ore processing on the recovery of REEs but also a perspective for commercial recovery of REEs in which both the grades of REEs and their occurrences (REE-bearing phases and REEs distributions) are very important parameters.

For different phosphate ores throughout the world the contents of total REEs vary from 0.03 to 1.0 wt. % [1,2]. This study was based on a phosphate ore with the content of total REEs 0.05 wt. %. As most apatite ores have similar mineralogy of REEs the methods and outcomes from this study are significantly valuable for the studies of REEs recovery from other apatite ores.

According to this study, the beneficiation tailings and PG are certainly the main sources of REEs in this apatite ore processing. They contain 26% and 62% of the REEs from the rock, respectively. In the tailings REEs occur in the phases of apatite, monazite, allanite and pyrochlore. Flotation could be used to concentrate apatite and monazite using fatty acids or hydroxamates as the collectors but allanite and pyrochlore have normally poor floatability using conventional reagents. Wet high intensity or gradient magnetic separation could improve their recoveries according to our previous study [29]. Fine grain sizes of the REE-bearing phases, low degrees of their liberation and low REEs grade would cause the recovery of REEs very challenging by physical techniques (flotation and magnetic separation). Limited studies have been conducted on REE recovery from the beneficiation tailings. In a study [3] of a phosphate ore from Florida, USA the process of gravity separation using a shaking table followed by flotation was applied to concentrate the REE-bearing minerals, monazite and xenotime. Then sulfuric acid leaching was used to recover the REEs from the concentrate but the REE recovery of the physical process was about 30% and the final recovery of the REE was less than 20%.

In the PG, the REEs mainly occur in the phases of gypsum and monazite. Physical methods such as screening, flotation and magnetic separation could be applied to concentrate monazite but high recovery of REEs is not expected because only 26% of the REEs are carried by monazite in the studied sample. Most (73% REOs) REEs carried by gypsum cannot be enriched by physical methods. In addition, higher contents of REEs are distributed in very fine size fractions, such as in the fine fraction of −45 μm as showed in Table 5. Physical processes become more challenging for the concentration of REEs from the PG.

Author Contributions: Conceptualization, investigation, writing—original draft preparation and funding acquisition, X.Y.; mineralogical studies and writing—review and editing, H.T.M. and L.P.

Funding: The authors are grateful to the Academy of Finland (298094) for the financial support.

Conflicts of Interest: The authors declare no conflict of interest.

References

1. Binnemans, K.; Jones, P.T.; Blanpain, B.; Van Gerven, T.; Pontikes, Y. Towards zero-waste valorisation of rare-earth-containing industrial process residues: A critical review. *J. Clean. Prod.* **2015**, *99*, 17–38. [CrossRef]
2. Grabas, K.; Pawelczyk, A.; Strek, W.; Szełęg, E.; Stręk, S. Study on the properties of waste apatite phosphogypsum as a raw material of prospective applications. *Waste Biomass Valor.* **2018**, 1–13. [CrossRef]
3. Zhang, P.; Liang, H.; DePaoli, D. Recovery of rare earths and P from a phosphate flotation tails. In Proceedings of the IMPC 2018 Physical Enrichment—Gravity, Magnetic and Electrostatic Separation, Moscow, Russia, 17–21 September 2018. Paper 761.
4. Hammas-Nasri, I.; Horchani-Naifer, K.; Férid, M.; Barca, D. Rare earths concentration from phosphogypsum waste by two-step leaching method. *Int. J. Miner. Process.* **2016**, *149*, 78–83. [CrossRef]
5. Jarosiński, A. Development of the Polish wasteless technology of apatite phosphogypsum utilization with recovery of rare earths. *J. Alloy. Compd.* **1993**, *200*, 147–150. [CrossRef]
6. Walawalkar, M.; Nichol, C.K.; Azimi, G. Process investigation of the acid leaching of rare earth elements from phosphogypsum using HCl, HNO_3 and H_2SO_4. *Hydrometallurgy* **2016**, *166*, 195–204. [CrossRef]
7. Azimi, G.; Papangelakis, V.G.; Dutrizac, J.E. Modelling of calcium sulphate solubility in concentrated multi-component sulphate solutions. *Fluid Phase Equilibria* **2007**, *260*, 300–315. [CrossRef]
8. Azimi, G.; Papangelakis, V.G.; Dutrizac, J.E. Development of an MSE-based chemical model for the solubility of calcium sulphate in mixed chloride–sulphate solutions. *Fluid Phase Equilibria* **2008**, *266*, 172–186. [CrossRef]
9. Ismail, Z.H.; Abu Elgoud, E.M.; Abdel Hai, F.; Ali, I.O.; Gasser, M.S.; Aly, H.F. Leaching of Some Lanthanides from Phosphogypsum Fertilizers by Mineral Acids. *Arab J. Nucl. Sci. Appl.* **2015**, *48*, 37–50.
10. Todorovsky, D.; Terziev, A.; Milanova, M. Influence of mechanoactivation on rare earths leaching from phosphogypsum. *Hydrometallurgy* **1997**, *45*, 13–19. [CrossRef]
11. Reid, S.; Walawalkar, M.; Azimi, G. Valorization of rare earth-containing landfilled stocks of industrial process residues: Phosphogypsum and red mud. In Proceedings of the Eres 2017, Santorini, Greece, 28–31 May 2017.
12. Preston, J.S.; Cole, P.M.; Craig, W.M.; Feather, A.M. The recovery of rare earth oxides from a phosphoric acid by-product. The recovery of rare earth oxides from a phosphoric acid by-product. Part 1: Leaching of rare earth values and recovery of a mixed rare earth oxide by solvent extraction. *Hydrometallurgy* **1996**, *41*, 1–19. [CrossRef]
13. Rutherford, P.M.; Dudas, M.J.; Samek, R.A. Environmental impacts of phosphogypsum. *Sci. Total Environ.* **1994**, *149*, 1–38. [CrossRef]
14. Dutrizac, J.E. The behaviour of the rare earth elements during gypsum ($CaSO_4 \cdot 2H_2O$) precipitation. *Hydrometallurgy* **2017**, *174*, 38–46. [CrossRef]
15. Harbi, H.M.; Eldougdoug, A.A.; El-Shahwai, M.S. Mineral processing and extraction of rare earth elements from the Wadi Khamal nelsonite ore, northwestern Saudi Arabia. *Arab. J. Geosci.* **2011**, *4*, 353–363. [CrossRef]
16. Koopman, C.; Witkamp, G.J. Extraction of lanthanides from the phosphoric acid production process to gain a purified gypsum and a valuable lanthanide by-product. *Hydrometallurgy* **2000**, *58*, 51–60. [CrossRef]
17. Ogawa, Y.; Shikazono, N. REE behavior during anhydrite and gypsum formations of the Kuroko type massive sulfide-sulfate deposits. In Proceedings of the 3rd International Workshop on Water Dynamics, Sendai Japan, 16–17 November 2005; pp. 162–166.

18. Ogawa, Y.; Shikazono, N.; Ishiyama, D.; Sato, H.; Mizuta, T.; Nakano, T. Mechanisms for anhydrite and gypsum formation in the Kuroko massive sulfidesulfate deposits, north Japan. *Mineral. Depos.* **2007**, *42*, 219–233. [CrossRef]
19. Sandström, A.; Fredriksson, A. Apatite for extraction—Leaching of Kiirunavaara apatite for simultaneous production of fertilizers and REE. In Proceedings of the XXVI International Mineral Processing Congress—IMPC-2012, New Delhi, India, 24–28 September 2012; pp. 4707–4714.
20. Wang, L.; Long, Z.; Huang, X.; Yu, Y.; Cui, D.; Zhang, G. Recovery of rare earths from wet-process phosphoric acid. *Hydrometallurgy* **2010**, *101*, 41–47. [CrossRef]
21. Jin, H.X.; Li, J.Q.; Wu, F.Z. Acidolysis kinetics and RE leaching mechanics of RE bearing phosphorite ores. *J. Univ. Sci. Technol. Beijing* **2011**, *33*, 1071–1078.
22. Santos, A.J.G.; Mazzilli, B.P.; Fávaro, D.I.T.; Sliva, P.S.C. Partitioning of radionuclides and trace elements in phosphogypsum and its source materials based on sequential extraction methods. *J. Environ. Radioact.* **2006**, *87*, 52–61. [CrossRef]
23. Borges, R.C.; Fávaro, D.I.T.; Caldas, V.G.; Lauria, D.C.; Bernedo, A.V.B. Instrumental neutron activation analysis, gamma spectrometry and geographic information system techniques in the determination and mapping of rare earth element in phosphogypsum stacks. *Environ. Earth Sci.* **2016**, *75*, 705. [CrossRef]
24. Schmidt, M.; Stumpf, T.; Walther, C.; Geckeis, H.; Franghänel, T. Incorporation versus adsorption: Substitution of Ca^{2+} by Eu^{3+} and Cm^{3+} in aragonite and gypsum. *Dalton Trans.* **2009**, *7*, 6645–6650. [CrossRef]
25. Alhassanieh, O.; Mrad, O.; Ajji, Z. Sorption and migration of Cs, Sr and Eu in gypsum-groundwater system. *Nukleonika* **2012**, *57*, 125–131.
26. Howie, R.A.; Zussman, J.; Deer, W. *Introduction to the Rock-Forming Minerals*, 3rd ed.; Mineralogical Society: Chantilly, VA, USA, 2013.
27. Pouchou, J.L.; Pichoir, F. Basic expression of "PAP" computation for quantitative EPMA. In Proceedings of the 11 the International Congress on X-Ray Optics and Microanalysis (ICXOM), London, ON, Canada, 4–8 August 1986; pp. 249–253.
28. Kalashnikov, A.; Yakovenchuk, V.N.; Pakhomovsky, Y.A.; Bazai, A.V.; Sokharev, V.A.; Konopleva, N.G.; Mikhailova, J.A.; Goryainov, P.M.; Ivanyuk, G.Y. Scandium of the Kovdor magnetite-apatite-baddeleyite deposit (Murmansk Region, Russia): Mineralogy, spatial distribution and potential resource. *Ore Geol. Rev.* **2016**, *72*, 532–537.
29. Yang, X.; Satur, J.V.; Sanematsu, K.; Laukkanen, J.; Saastamoinen, T. Beneficiation studies of a complex REE ore. *Miner. Eng.* **2015**, *71*, 55–64. [CrossRef]

© 2019 by the authors. Licensee MDPI, Basel, Switzerland. This article is an open access article distributed under the terms and conditions of the Creative Commons Attribution (CC BY) license (http://creativecommons.org/licenses/by/4.0/).

Article

Geochemical Characteristics of Dolomitic Phosphorite Containing Rare Earth Elements and Its Weathered Ore

Shuai Li [1], Jie Zhang [2], Huaifa Wang [1,*] and Caili Wang [1]

1. College of Mining Engineering, Taiyuan University of Technology, Taiyuan 030024, China
2. Mining College, Guizhou University, Guiyang 550025, China
* Correspondence: wanghuaifa@tyut.edu.cn

Received: 12 April 2019; Accepted: 4 July 2019; Published: 8 July 2019

Abstract: In order to provide a good theoretical guidance for the development and utilization of weathered phosphorite resources, we investigated the geochemical and mineralogical characteristics of primary and weathered phosphorites. The analysis of trace elements showed that the primary ore has hydrothermal sedimentation effect in the later stage, the weathered ore has obvious residual enrichment and the phosphate ore belongs to clastic lithologic phosphate rock. In addition, through leaching test method, it was shown that rare earth elements are present in fluorapatite in the form of isomorphic substitution, and the proportion of rare earth elements adsorbed on clay and other minerals was likely to be between 2% and 3%. The light rare earth elements are relatively enriched in both primary and weathered phosphorite, and Ce and Eu have obvious negative anomalies. The primary phosphorite is a dolomitic phosphorite containing rare earth elements, which are naturally enriched by weathering, and its weathered ore has obvious residual enrichment, while the deposit was characterized by normal marine sedimentation and hydrothermal action.

Keywords: phosphorite; weathered ore; geochemical characteristic; rare earth elements; technological mineralogy

1. Introduction

Phosphorite is a non-renewable resource without substitute and a significant raw material for phosphatic fertilizers and phosphorus-based chemicals [1–3]. There are abundant sedimentary phosphorite deposits in China, most of which were formed in the Neoproterozoic Late Sinian and in the Paleozoic early Cambrian [4,5]. Supergene weathering of these phosphorites produced weathered phosphorite, a potential industrial type of phosphorite, which is a high-quality ore [6]. The formation of weathered phosphorite is restricted and controlled by the geochemical characteristics of primary ore and the external conditions of ore bodies or rock masses presence [7]. Weathering changed the characteristics of phosphorite, and there is a significant difference between weathered ore and primary ore, regardless of their characteristics or grade [8]. Relative to fresh phosphorite, weathered phosphorite is generally enriched in alkali metals, alkaline earth elements, halogen and radioactive elements, and depleted in mantle elements [7,9]. Weathered phosphorites are also characterized by generally enrichment in light rare earth elements (LREE), depletion in HREE, and Eu negative anomaly [10,11].

Different rocks and ores have different resistance to weathering. Carbonate is easier to be weathered and denuded than phosphate, and the weathering products, calcium, and magnesium are easily leached away, residually enriching apatite [12–14]. The phosphorites of Gezhongwu Formation in Zhijin County mainly are bioclast dolomitic and siliceous. The bioclastic components of the phosphorites consist of pelecypods, gastropods, algae, etc [15]. Wang et al. (2004) studied the genesis of the Xinhua phosphorite deposit in Guizhou China and analyzed the REE composition of the ore using ICP-MS [16]. They concluded that phosphorite was formed exclusively by marine sediments deposited in ancient pericontinental seas. They also suggested that the phosphorite deposit was subjected to strong weathering after deep buried diagenetic processes, which did not change its REE composition remarkably. A systematic study of leaching of rare earths from a phosphorite ore using four leaching modes [17]: sulfuric acid leaching, phosphoric acid leaching, mixed acid leaching, and two-step leaching of phosphoric acid and sulfuric acid was carried out by Jin et al.(2017). Wang (2011) found that the Sr/Ba ratio of weathered phosphorite ores is generally less than 1 due to dissolution of dolomite [7], and the U/Th ratio is generally significantly lower than that of primary phosphorite ores.

Although some geological work has been done on the phosphorite deposits in southern China, the weathered phosphorite has not been fully recognized and studied. In this study, we investigated the mineralogical and geochemical composition of the primary and weathered phosphorites to optimize the use of weathered phosphorite and their rare earth elements in industry.

2. Materials and Methods

2.1. Materials

Six phosphorite samples from different locations in the Zhijin mining district in the Guizhou Province, China, were analyzed for their major oxides and trace elements composition, using the lithum borate/lithum metaborate and X spectrofluorimetry methods [18,19]. The assay results are summarized in Table 1. The samples 1 and 2 are form primary phosphorites and samples 3 to 6 are form weathered phosphorites according to the weathered phosphorite criteria proposed by Zhang et al. [6,20].

Table 1. Major oxides (%) and trace elements(ppm) of the studied phosphorite samples.

Oxides	Phosphorite					
	Primary		Weathered			
	Sample No.					
	1	2	3	4	5	6
P_2O_5	19.36	9.42	30.12	29.21	31.01	35.18
SiO_2	13.61	24.95	17.59	18.29	17.69	10.15
Al_2O_3	1.57	1.71	4.25	4.75	2.91	2.23
Fe_2O_3	0.99	0.95	3.28	3.49	2.74	1.98
CaO	35.75	26.98	38.73	37.59	39.51	44.72
MgO	7.79	10.36	0.55	0.61	0.54	0.27
Na_2O	0.06	0.07	0.08	0.08	0.06	0.09
K_2O	0.42	0.53	1.13	1.27	0.90	0.57
TiO_2	0.08	0.05	0.12	0.13	0.12	0.11
L.O.I	18.50	23.70	3.90	4.06	3.33	2.78
Total	98.13	98.72	99.75	99.48	98.81	98.08
Trace Elements (ppm)						
Ba	183	3200	433	479	732	455
Ce	142	78	247	232	210	272
Cr	20	20	40	40	50	40
Cs	1	1	2	2	2	1
Dy	27	14	46	44	42	53
Er	15	8	26	25	24	29
Eu	6	4	10	10	13	14
Ga	4	3	9	10	7	6
Gd	34	18	58	55	53	69
Hf	1	1	2	2	1	2
Ho	6	3	10	9	9	11
La	207	103	370	353	288	406
Lu	1	1	2	2	2	2
Nb	2	2	4	4	2	2
Nd	150	81	258	241	227	307
Pr	35	19	62	58	52	72

Table 1. Cont.

	Trace Elements (ppm)					
Rb	9	8	27	28	19	14
Sm	27	14	45	42	41	54
Sn	1	1	1	1	1	1
Sr	679	335	677	722	698	1305
Tb	5	2	8	7	7	9
Th	7	5	9	9	5	9
Tm	2	1	3	3	3	3
U	5	3	10	9	18	11
V	19	8	33	32	72	24
W	2	2	2	2	1	3
Y	317	166	566	538	518	602
Yb	8	5	14	14	13	16
Zr	49	37	74	66	44	54
U/Th	0.64	0.57	1.03	1.01	3.55	1.11
Sr/Ba	3.72	0.10	1.56	1.51	0.95	2.87
Rb/Cs	14.92	10.25	15.46	14.71	10.32	16.63
Zr/Hf	37.69	37.00	35.24	31.43	33.85	31.76

2.2. Methods

2.2.1. X-ray Diffraction (XRD) Analysis

In order to identify the mineral composition of phosphorite samples, an X-ray powder diffract meter (Rigaku D/MAX 2500, Tokyo, Japan) was applied by using Cu Kα radiation (40 kV, 100 mA). The samples were scanned at the speed of 6°/min over a 2θ range of 2.6–45°. Furthermore, SIROQUANT software (V3, Sietronics) was carried out to quantify the mineralogical composition. The software used the full-profile Rietveld method to refine the shape of calculated XRD pattern against the profile of a measured pattern [21].

2.2.2. Optical Microscopy

Thin section of phosphorite samples were prepared and investigate by optical microscopy [22,23]. An Olympus CX21 polarizing microscope was used to investigate the mineral characteristics and textural relationships of phosphorite.

2.2.3. SEM-EDX

The morphology, composition, and orientation of phosphorites were studied using a Hitachi 3400 N scanning electron microscope (SEM) in Guizhou University Physical and Chemical Testing Center lab. The morphology was studied using back scattered electrons (BSE) and the composition of the various phases using EDX [24–26].

2.3. Leaching Test

In order to determine the presence of REE, phosphorite samples were leached with different acids and inorganic salts [17]. A 10 g composite sample was leached with acid (HCl or HNO$_3$) and whereas a 5 g composite samples was leached with inorganic salt. The experiments were carried out at room temperature for 2 h in 100 mL leaching agent with acids concentrations of 1%, 3%, 5%, 7%, 9%, and 11% and inorganic salts concentrations of 5%, 10%, 15%, 20%, and 25%. The main reactions can be summarized as follows [27]:

$$Ca_5(PO_4)_3F + 10HCl \rightarrow 3H_3PO_4 + 5CaCl_2 + HF(g) \tag{1}$$

$$Ca_5(PO_4)_3F + 10HNO_3 \rightarrow 3H_3PO_4 + 5Ca(NO_3)_2 + HF(g) \tag{2}$$

$$2RE^{3+} + Ca_5(PO_4)_3F \rightarrow (2RE^{3+}, 3Ca^{2+})[PO_4]F + 2Ca^{2+} \tag{3}$$

The four leaching agents were hydrochloric acid (HCl), nitric acid (HNO$_3$), ammonium chloride (NH$_4$Cl) and sodium chloride (NaCl). HCl and HNO$_3$, both with an analytical purity, were purchased from Chongqing Chuandong Chemical (Group) Co., Ltd. (Chongqing, China); while the other reagents were obtained from Tianjin Kemiou Chemical Reagent Co., Ltd. (Tianjin, China).

2.4. Mathematical Method for Judging Independent Minerals

Independent minerals, concentrated state of elements, are particles larger than 0.001 mm in diameter and can be studied with naked eyes or under the microscope. There are two basic conditions for the formation of independent minerals [17]: one is relative stability under certain physical and chemical conditions; the other is certain element content. It is necessary to carry out mathematical analysis on the content of phosphorus and rare earth elements in the phosphate rock and calculate their mean value and mean variance. If the ratio of the mean value and mean variance of the two are significantly different (generally considered to be 20%), rare earth will exist in the form of independent mineral; otherwise, it will exist in a dispersed form such as homogeneity, adsorption, etc.

When a dependent variable, y, varies to some extent with an independent variable, x, it can be assumed that they have the following relationship: $y = a + bx$. According to the analysis in Table 1, we let y be the content of rare earth elements and x be the content of phosphorus pentoxide. The regression equation was established, and the coefficients a and b were obtained using Equations (4–8).

$$b = \frac{\sum (x_i - \bar{x})(y_i - \bar{y})}{\sum (x_i - \bar{x})^2} \quad (4)$$

$$a = \bar{y} - b\bar{x} \quad (5)$$

$$\bar{x} = \frac{1}{N_{sa}} \sum x_i \quad (6)$$

$$s_x = \sqrt{\frac{\sum (x_i - \bar{x})^2}{N_{sa} - 1}} \quad (7)$$

$$s_y = \sqrt{\frac{\sum (y_i - \bar{y})^2}{N_{sa} - 1}} \quad (8)$$

where $\bar{y}$ and $\bar{x}$ are the mean values of dependent variable y and independent variable x, respectively; s_x and s_y are the mean square deviation of x and y, respectively; N_{sa} is total number of samples.

3. Results and Discussion

3.1. Comparion of Mineral Composition between Primary Weathered Phosphorites

All the samples are rich in CaO, P$_2$O$_5$ and SiO$_2$ (Table 1). The L.O.I in primary phosphorite is higher than in weathered ore. Fluorapatite is present in both primary and weathered ores, with respective averages of 31.7% and 79.3% (Figure 1). Dolomite was detected only in primary ore with an average content of 48.75% (Figure 1 no. 1 and 2).

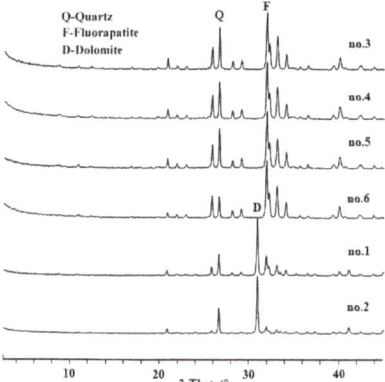

Figure 1. X-ray diffraction (XRD) patterns of studied phosphorite samples.

3.2. Ore Texture Differences

Figure 2 shows the distribution and textural characteristics of various minerals present in the ore. The composition of the various minerals was studied with a SEM equipped with an EDX detector (Figure 3) [28,29]. The results showed that fluorapatite, embedded in dolomitic matrix, is the main phosphorus-bearing mineral. Fluorapatite occurs in spherical, oolitic, and clastic forms, in mainly grey-yellow and grey-brown colors under transmitted light. Dolomite is the main cement in the rock and is colorless or white under plane parallel light. Dolomite also appears in the matrix of phosphorite as euhedral and subhedral isolated crystals and aggregate. Quartz was randomly distributed in the dolomite matrix, with a few crystals disseminated in phosphorite. It occurs as granular and hypidiomorphic grains with clean surfaces.

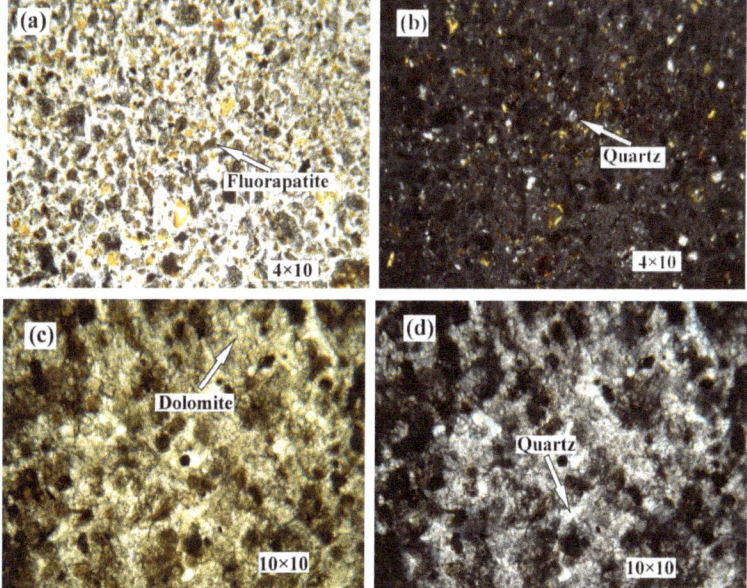

Figure 2. *Cont.*

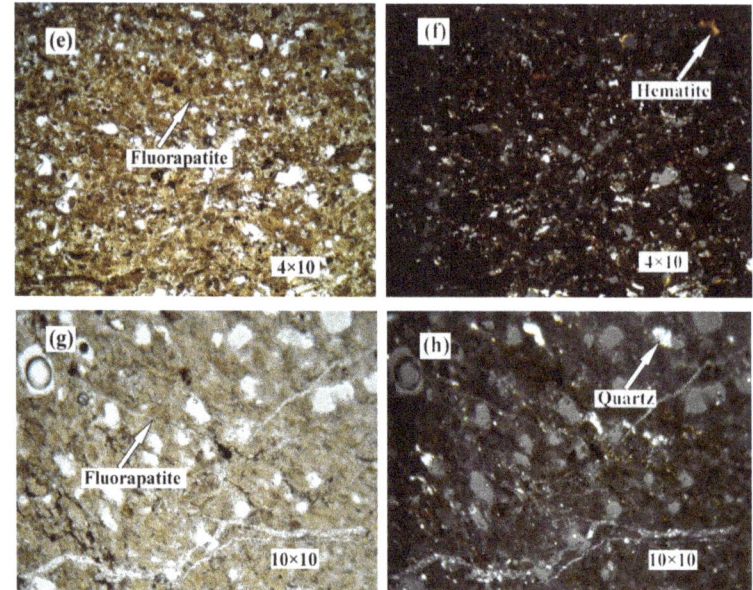

Figure 2. Photomicrographs of phosphorite. (**a**) no.1 crossed nicols. (**b**) no.1 perpendicular polarized light. (**c**) no.2 plane polarized light. (**d**) no.2 crossed nicols. (**e**) no.4 plane polarized light. (**f**) no.4 crossed nicols. (**g**) no.5 plane polarized light. (**h**) no.5 crossed nicols.

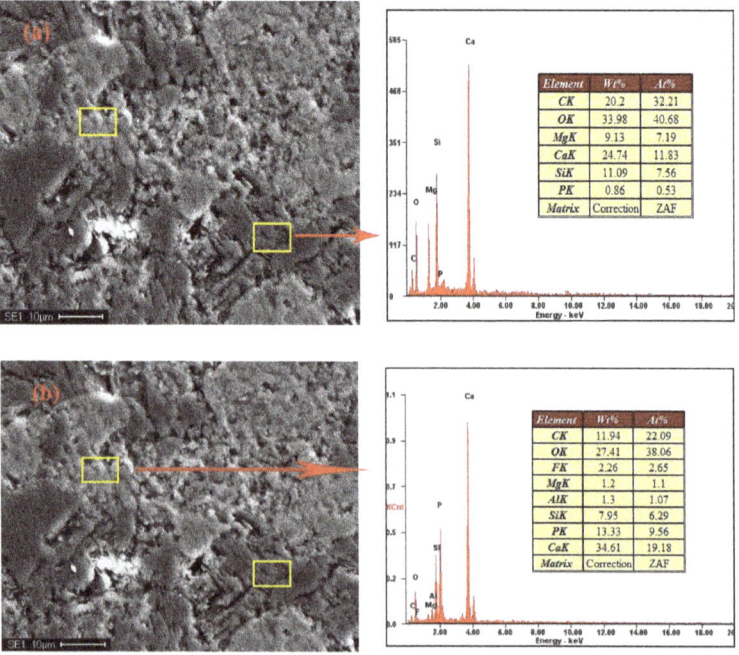

Figure 3. *Cont.*

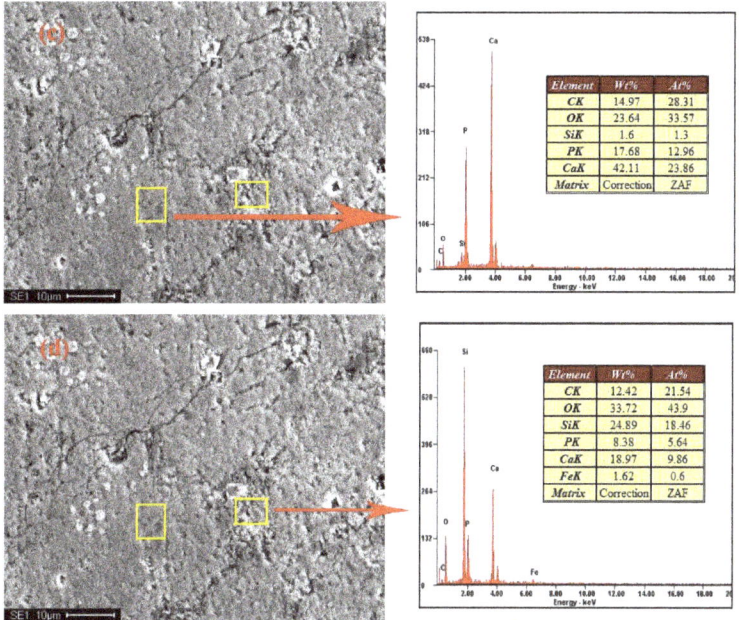

Figure 3. SEM-EDX results of samples. (**a**,**b**) no.1; (**c**,**d**) no.4.

3.3. Comparison of Trace Elements

As shown in Table 1, the primary and weathered ores are enriched in Ba, Ce, La, Nd, Sr, and Y, and depleted in Co and Ni. Gallium, a chalcophile element related to hydrothermal activity [22], was enriched in the weathered ore and gradually depleted with the increase in temperature. Chromium, a siderophile element mainly found in ultrabasic and basic rocks, was also enriched in the weathered ore and mainly found in basic rocks and ultrabasic rocks.

U and Th are lithophile elements with similar geochemical properties, and they usually occur in oxides or oxygen-bearing salts [19]. U/Th values (see Table 1) of primary ore are lower than 1, while those of weathered ore are higher than 1, which indicates that hydrothermal precipitation has an effect on the later stage of primary ore, and weathered ore has obvious residual enrichment properties. The average Sr/Ba values (Table 1) of primary and weathered ores are all greater than 1 except for sample No.6, indicating that the phosphate rock is clastic in origin. The mean Rb/Cs values of primary ore and weathered ore were almost the same, indicating that these active lithophile elements were adsorbed or bound evenly in the minerals of the original rock. Both Zr and Hf are high field strength elements and indicate differences in hydrothermal action, the former was enriched at an early stage while the latter at a later stage. The Zr/Hf value of weathered ore was significantly smaller than that of primary ore, indicating that primary ore was related to the early hydrothermal activity while weathered ore was relevant to the later stage.

3.4. Comparison of Rare Earth Elements

As shown in Table 2, the content of total rare earth elements in weathered phosphate rock was higher than that in primary ore. Concentration of LREE is higher than that of HREE in both types of ore (average LREE/HREE ratio = 1.35, average (La/Yb)$_N$ ratio = 16.11), and δCe and δEu are obtained using Equations (9) and (10). Figure 4 shows negative Ce and Eu anomalies in primary and weathered ores, while La, Nd and Y are relatively enriched. The negative Ce anomaly indicates that there may be two situations: one is that the ore-forming environment of phosphate rock was oxidizing and Ce^{4+} with small ion radius was difficult to enter apatite lattice to form a loss; the other is that Ce^{4+} can easily form CeO_2 which lead to loss during the reworking of phosphate rock in marine environment. The negative Eu anomalies indicate that magma was differentiated by a certain intensity of crystallization [14–17]. In summary, the characteristics of REE indicated that the phosphorite belongs to the normal marine sedimentary rock.

$$\delta Ce = Ce_N/(La_N \times Pr_N)^{0.5} \tag{9}$$

$$\delta Eu = Eu_N/(Sm_N \times Gd_N)^{0.5} \tag{10}$$

Table 2. Rare earth elements(ppm) of phosphorite.

REE	Primary Phosphorite			Weathered Phosphorite		
	Sample No.					
	1	2	3	4	5	6
La	207	102.5	370	353	288	406
Ce	142	78.3	247	232	210	272
Pr	35.4	19.1	61.6	58.1	52	72.1
Nd	149.5	80.6	258	241	227	307
Sm	26.6	14.5	45.4	42.2	41.1	54.3
Eu	6.34	3.65	10.45	9.89	12.6	14
Gd	33.8	18.45	57.6	54.8	52.8	69.1
Tb	4.55	2.48	7.78	7.35	7.02	9.05
Dy	27	14.3	46.2	43.7	41.7	53
Ho	5.63	3.04	9.86	9.33	9	11.25
Er	14.65	8.05	26	24.7	23.7	29.1
Tm	1.71	0.97	2.98	2.85	2.69	3.28
Yb	8.43	4.76	14.35	13.75	12.8	16.25
Lu	1.13	0.67	1.87	1.81	1.65	2.08
Y	317	165.5	566	538	518	602
ΣREE	980.74	516.82	1725.09	1632.48	1500.06	1920.51
LREE	566.84	298.6	992.45	936.19	830.7	1125.4
HREE	413.9	218.22	732.64	696.29	669.36	795.11
LREE/HREE	1.37	1.37	1.35	1.34	1.24	1.42
$(La/Yb)_N$	16.37	14.36	17.19	17.12	15	16.66
δEu	0.64	0.68	0.62	0.63	0.82	0.70
δCe	0.36	0.39	0.35	0.35	0.38	0.35

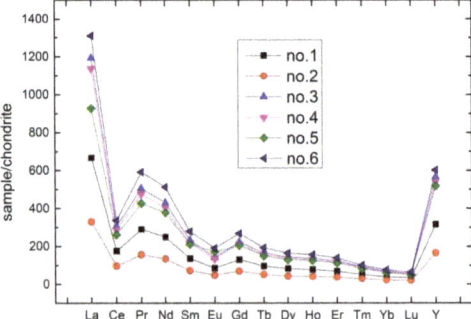

Figure 4. Normalized rare earth elements (REE) patterns of phosphorite samples (after Boynton, 1984 [30]).

3.5. Analysis of Rare Earth Elements (REE) Occurrence State

3.5.1. Analysis of Independent Form

After calculation, the average values $\bar{x} = 25.7170$ and $\bar{y} = 0.1379$, the mean square errors $s_x = 9.5433$ and $s_y = 0.0528$, and the regression equation $y = 0.0054x - 0.0019$ were obtained. As a consequence, the values $\left(\frac{\bar{x}}{\bar{y}}\right) = 186.451$ and $\left(\frac{s_x}{s_y}\right) = 180.747$ were also acquired, and they differ by 3.06%, which is much less than 20%. Therefore, it is considered that the rare earth elements are unlikely to exist as an independent mineral [15,31,32].

3.5.2. Analysis of Isomorphic Form

Rare earth elements can either exist in the form of independent minerals or isomorphic substitution in the minerals' lattices. If the REE exists in the form of a separate mineral such as monazite and xenotime, it is difficult to dissolve in the dilute acid, even in small amounts. If the rare earth elements exist in the crystal lattice of colloidal phosphate ore, their physicochemical properties should be similar, the rare earth elements can be completely dissolved in dilute acid together with the colloidal phosphate [31]. Therefore, under certain conditions, the fluorapatite in phosphorite can be dissolved by dilute acid, and the leaching rate of phosphorus and rare earth elements can be analyzed to study the deportment of rare earth elements [31].

The leaching process (see Equations (1)–(3)) was carried out using a magnetic stirrer at room temperature. It can be seen from Figure 5 that the leaching rates of phosphorus and rare earth are approximately the same. When the leaching rate reaches 99.98%, it no longer increases with increase of acid concentration, meaning they exist in the form of isomorphism. With the dissolution and lattice destruction of apatite, phosphorus and rare earth elements were transferred into the solution at the same time. The isomorphic substitution involves the replacement of Ca^{2+} in apatite by REE [31].

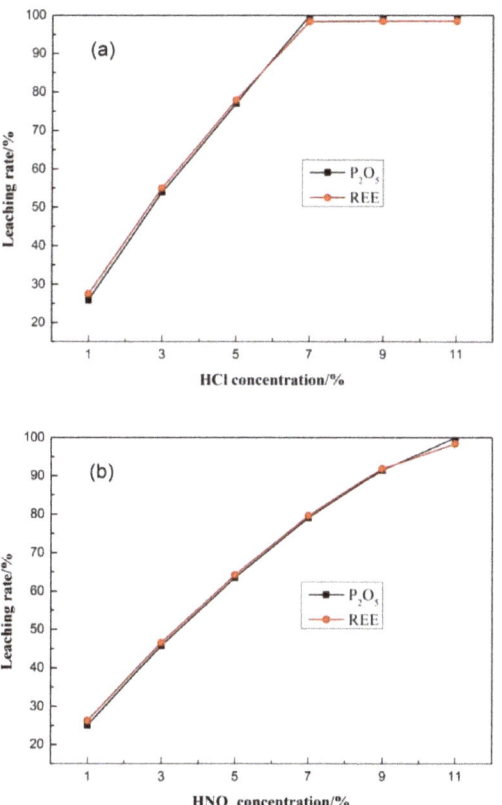

Figure 5. Results of acid leaching test. (**a**) HCl leaching test; (**b**)HNO$_3$ leaching test.

3.5.3. Analysis of Ion Adsorption Form

If rare earth elements are adsorbed on clays or other minerals, they (REE) can generally be leached out by inorganic salts [33–35]. In this experiment, a 5 g composite sample was leached with NH_4Cl and NaCl for 2 h in 100 mL inorganic salts with different concentrations and the results are shown in Figure 6. The leaching rate of rare earth increases slowly with increasing concentration of inorganic salts until at 2.6%, which indicates a low amount of adsorbed REE. Taking into account experimental errors, its content is likely to be between 2% and 3% [33–36].

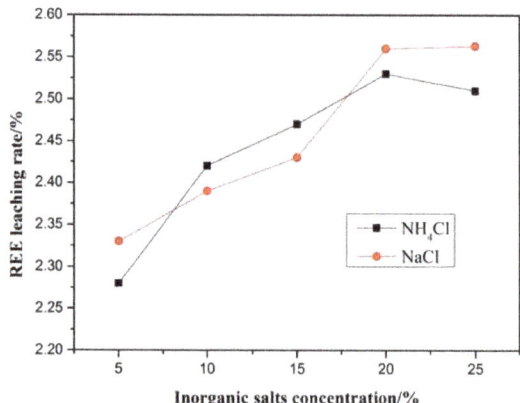

Figure 6. Results of inorganic salts leaching test.

4. Conclusions

The weathering of primary phosphorite in southern China increases the concentration of P_2O_5, Fe_2O_3, and Al_2O_3, and decreases the MgO, CaO, and CO_2 content. The concentration of trace elements including Ba, Ce, La, Nd, Sr, and Y was also increased, while the Co and Ni content was strongly depleted, indicating the inheritance from the original rocks. The U/Th ratios indicate that the primary ore was affected by hydrothermal imprint in the later stage of sedimentation, and the weathered ore was residually enriched in REE. Light rare earth elements are enriched relative to HREE in both primary and weathered ores, and Ce and Eu both have negative anomalies. The REE occurs mainly in fluorapatite crystal lattice and in isomorphic substitution with Ca, with only a small proportion (2–3%) adsorbed onto the mineral surface. Their ability to form separate minerals is low. The characteristics of rare earth elements indicated that the phosphorite deposits are normal marine sedimentary rocks.

Author Contributions: The contributions of the authors were as follows: conceptualization, J.Z. and S.L.; methodology, J.Z. and S.L.; validation, S.L., J.Z., H.W. and C.W.; investigation, J.Z. and S.L.; resources, J.Z. and S.L.; data curation, S.L.; writing—original draft preparation, S.L.; writing—review and editing, S.L.; visualization, S.L.; supervision, C.W.; project administration, J.Z.; funding acquisition, H.W.

Funding: This research was funded by the National Natural Science Foundation of China, grant number 51164004, and the APC was funded by Huaifa Wang.

Acknowledgments: Thanks for the software and formal analysis provided by Aoshi Analytical Testing (Guangzhou, China) Co., Ltd.

Conflicts of Interest: The authors declare no conflict of interest.

References

1. Abouzeid, A.Z.A. Physical and thermal treatment of phosphate ores—An overview. *Int. J. Miner. Process.* **2008**, *85*, 59–84. [CrossRef]
2. dos Santos, M.A.; Santana, R.C.; Capponi, F.; Ataide, C.H.; Barrozo, M.A.S. Effect of ionic species on the performance of apatite flotation. *Sep. Purif. Technol.* **2010**, *76*, 15–20. [CrossRef]

3. Gharabaghi, M.; Irannajad, M.; Noaparast, M. A review of the beneficiation of calcareous phosphate ores using organic acid leaching. *Hydrometallurgy* **2010**, *103*, 96–107. [CrossRef]
4. Lin, Y. *The Construction and Application of Phosphorus Resource Development and Utilization Model in China*; Hefei University of Tecnology: Hefei, China, 2010.
5. Zhang, W.; Ma, W.; Zhang, F.; Ma, J. Comparative analysis of the superiority of China's phophate rock and development strategies with that of the United States and Morocco. *J. Nat. Resour.* **2005**, *20*, 378–386.
6. Li, T.; Wen, S.; Tian, P.; Tang, Z.; Huang, Y. Weathered phosphorites in south China. *Geol. Chem. Miner.* **1996**, *3*, 53–57.
7. Wang, J. *Study on Geology and Geochemistry of Wheathered Phosphate Deposits, South China*; China University of Geosciences: Beijing, China, 2011.
8. Chang, S.; Zhu, J.; Liu, Y.; He, J. Study status and trends of weathered phosphate rocks in Yunnan, Guizhou. *Ind. Miner. Process.* **2010**, *39*, 41–44.
9. Tian, S.; Wang, Q.; Zhu, H.; Zhu, Y.; Zhou, J.; Wang, J. Fixed quantity exploration and assessment of weathered phosphoric ore in southwest of China. *Geol. Chem. Miner.* **2004**, *4*, 205–209.
10. Wei, D.; Cui, X. Study on process characteristic of heavy rare earths in phosphorite in Guizhou province. *Conserv. Util. Miner. Resour.* **2003**, *2*, 38–40.
11. Yang, R.D.; Gao, H.; Wang, Q.; Bao, M. REE enrichment in Early Cambrian Gezhongwu formation phosphorous rock series in Sanjia, Zhijun County, Guizhou Province, China. *J. Rare Earths* **2005**, *23*, 760–767.
12. Ross, J.; Gao, L.; Meouch, O.; Anthony, E.; Sutarwala, D.; Mamo, H. Carbonate Apatite Precipitation from Synthetic Municipal Wastewater. *Minerals* **2017**, *7*, 129. [CrossRef]
13. Salama, W.; Khirekesh, Z.; Amini, A.; Bafti, B.S. Diagenetic evolution of the upper Devonian phosphorites, Alborz for Mountain Range, northern Iran. *Sediment. Geol.* **2018**, *376*, 90–112. [CrossRef]
14. Zhang, J.; Shun, C.; Yang, G.; Xie, F. Separation and enrichment of rare earth elements in phosphorite in Xinhua, Zhijin, Guizhou. *J. Rare Earths* **2006**, *24*, 413.
15. Shi, C.; Hu, R.; Wang, G. Study on REE geochemistry of Zhijin phosphorites, Guizhou province. *J. Mineral. Petrol.* **2004**, *24*, 71–75.
16. Wang, M.; Sun, X.; Ma, M. Rare earth elements geochemistry and genesis of Xinhua large-size phosphorite deposit in western Guizhou. *Miner. Depos.* **2004**, *23*, 484–493.
17. Jin, H.-X.; Wu, F.-Z.; Mao, X.-H.; Wang, M.-L.; Xie, H.-Y. Leaching isomorphism rare earths from phosphorite ore by sulfuric acid and phosphoric acid. *Rare Met.* **2017**, *36*, 840–850. [CrossRef]
18. Lednev, A.V.; Lozhkin, A.V. Remediation of agrosoddy-podzolic soils contaminated with cadmium. *Eurasian Soil Sci.* **2017**, *50*, 620–629. [CrossRef]
19. Li, J.; Jin, H.; Chen, Y.; Mao, X. Rare earth elements in Zhijin phosphorite and distribution in two-stage flotation process. *J. Rare Earths* **2007**, *25*, 85–90.
20. Zhang, J. The weathered ore criterion of Dianchi phosphorite. *Yunnan Metall.* **1992**, 24–30.
21. Li, X.; Zhang, Q.; Hou, B.; Ye, J.; Mao, S.; Li, X. Flotation separation of quartz from collophane using an amine collector and its adsorption mechanisms. *Powder Technol.* **2017**, *318*, 224–229. [CrossRef]
22. Gao, P.; He, Z.; Li, S.; Lash, G.G.; Li, B.; Huang, B.; Yan, D. Volcanic and hydrothermal activities recorded in phosphate nodules from the Lower Cambrian Niutitang Formation black shales in South China. *Palaeogeogr. Palaeoclimatol. Palaeoecol.* **2018**, *505*, 381–397. [CrossRef]
23. Huggett, J.; Hooker, J.N.; Cartwright, J. Very early diagenesis in a calcareous, organic-rich mudrock from Jordan. *Arab. J. Geosci.* **2017**, *10*, 270. [CrossRef]
24. Breiland, A.A.; Flood, B.E.; Nikrad, J.; Bakarich, J.; Husman, M.; Rhee, T.; Jones, R.S.; Bailey, J.V. Polyphosphate-Accumulating Bacteria: Potential Contributors to Mineral Dissolution in the Oral Cavity. *Appl. Environ. Microbiol.* **2018**, *84*. [CrossRef] [PubMed]
25. Chen, J.; Yang, R.; Wei, H.; Gao, J. Rare earth element geochemistry of Cambrian phosphorites from the Yangtze Region. *J. Rare Earths* **2013**, *31*, 101–112. [CrossRef]
26. Gamez Vintaned, J.A.; Linan, E.; Navarro, D.; Zhuravlev, A.Y. The oldest Cambrian skeletal fossils of Spain (Cadenas Ibericas, Aragon). *Geol. Mag.* **2018**, *155*, 1465–1474. [CrossRef]
27. Walawalkar, M.; Nichol, C.K.; Azimi, G. Process investigation of the acid leaching of rare earth elements from phosphogypsum using HCl, HNO$_3$, and H$_2$SO$_4$. *Hydrometallurgy* **2016**, *166*, 195–204. [CrossRef]

28. Linsy, P.; Nath, B.N.; Mascarenhas-Pereira, M.B.L.; Vinitha, P.V.; Ray, D.; Babu, C.P. Benthic cycling of phosphorus in the Eastern Arabian Sea: Evidence of present day phosphogenesis. *Mar. Chem.* **2018**, *199*, 53–66. [CrossRef]
29. Sinirkaya, M.; Ozer, A.K.; Gulaboglu, M.S. Kinetics of dissolution of Mardin-Mazidagi (Turkey) phosphate ore in dilute phosphoric acid solutions. *Miner. Metall. Process.* **2010**, *27*, 110–115. [CrossRef]
30. Boynton, W.V. Cosmochemistry of the Rare Earth Elements. In *Rare Earth Element Geochemistry*; Henderson, P., Ed.; Elsevier: Amsterdam, The Netherlands, 1984.
31. Xie, F.; Zhang, T.A.; Dreisinger, D.; Doyle, F. A critical review on solvent extraction of rare earths from aqueous solutions. *Miner. Eng.* **2014**, *56*, 10–28. [CrossRef]
32. Zhang, J.; Zhang, Q.; Chen, D. REE geochemistry of the ore-bearing REE in Xinhua phosphorite, Zhijin, Guizhou. *J. Mineral. Petrol.* **2003**, *23*, 35–38.
33. Baghdady, A.R.; Howari, F.M.; Al-Wakeel, M.I. On the mineral characteristics and geochemistry of the Florida phosphate of Four Corners and Hardee County mines. *Sediment. Geol.* **2016**, *342*, 1–14. [CrossRef]
34. Chen, J.; Yang, R.; Zhang, J. Mode of occurrence of rare earth elements in posphorite in Zhijin county, Guizhou province, China. *Acta Mineral. Sin.* **2010**, *30*, 123–129.
35. Gall, Q.; Davis, W.J.; Lowe, D.G.; Dabros, Q. Diagenetic apatite character and in situ ion microprobe U-Pb age, Keeseville Formation, Potsdam Group, New York State. *Can. J. Earth Sci.* **2017**, *54*, 785–797. [CrossRef]
36. Huang, X.-W.; Long, Z.-Q.; Wang, L.-S.; Feng, Z.-Y. Technology development for rare earth cleaner hydrometallurgy in China. *Rare Met.* **2015**, *34*, 215–222. [CrossRef]

© 2019 by the authors. Licensee MDPI, Basel, Switzerland. This article is an open access article distributed under the terms and conditions of the Creative Commons Attribution (CC BY) license (http://creativecommons.org/licenses/by/4.0/).

Article

Rare Earth Element Recovery from Acidic Extracts of Florida Phosphate Mining Materials Using Chelating Polymer 1-Octadecene, Polymer with 2,5-Furandione, Sodium Salt

Joseph P. Laurino *, Jack Mustacato and Zachary J. Huba

Periodic Products Inc., Fort Lauderdale, FL 33315, USA
* Correspondence: jlaurino@periodicproducts.com; Tel: +1-954-764-7654

Received: 12 June 2019; Accepted: 28 July 2019; Published: 6 August 2019

Abstract: To meet the growing global demand for rare earth elements (REEs), nontraditional mining sources of these metals are being investigated. Phosphate ore and phosphate mining wastes have been identified as possible alternative sources to REEs. In this study, REEs were extracted from Florida phosphate mining materials using mineral and organic acids. The REEs were then recovered at high efficiencies using a chelating polymer, 1-octadecene, polymer with 2,5-furandione, sodium salt. At pH 1.5, the chelation polymer effectively bound nearly 100% of the rare earth elements extracted from the solids. Overall extraction and recovery yields were between 80% for gadolinium and 8% for praseodymium from amine tailings, between 70% for terbium and 7% for praseodymium from phosphogypsum, between 56% for scandium and 15% for praseodymium from phosphate rock, and between 77% for samarium and 31% for praseodymium from waste clay. These results suggest that this chelating polymer efficiently recovers rare earth elements from acidic extracts of phosphate mining waste products.

Keywords: phosphate; phosphogypsum; acid extraction; REE recovery

---

1. Introduction

Phosphate mining and fertilizer production is a vital global industry and is essential to global agriculture and food markets. One of the most common phosphate sources is fluorapatite ($Ca_5(PO_4)_3F$), with monazite (($Ce,La,Nd,Th)PO_4,SiO_4$) and xenotime (YPO_4) being phosphate-containing minerals commonly found with fluorapatite at around 1% by mass [1]. While monazite and xenotime are sources of phosphate, they are also an important source of elements known as rare earth elements (REEs). REEs are commonly substituted in place of calcium in the mineral structure. REEs are a group of 17 elements comprising the 15 lanthanide elements along with yttrium and scandium. Although not part of the lanthanide series of elements, yttrium and scandium are included because they have similar geochemical properties. The REEs are important elements because they are used in many technology- and energy-related applications: high-strength permanent magnets, lightweight alloys, electric motors, petrochemical refining, and fluorescent/optical displays [2].

During phosphate mining in Florida, the phosphate ore is purified and concentrated to attain phosphate rock, while the less concentrated portions, in the form of phosphatic clay and sand, are discarded. The phosphate is then extracted from the phosphate rock as liquid phosphoric acid, with the residual solid being $CaSO_4$ or phosphogypsum. Due to the presence of monazite and xenotime, Florida phosphate rock, phosphate clay, and phosphogypsum have been shown to contain significant levels of REEs [3–9]. It is estimated that 100,000 tons of REEs are co-mined with phosphate rock annually; these REEs become discarded as waste or remain in the phosphate-containing fertilizer or end product, making them a potential source for REEs [1].

While using phosphate rock and phosphate mining materials as a new source of REE metals is promising, there are technological hurdles that need to be addressed to make the process economically viable. Currently, the extraction and separation of REEs involves complex chemistries as well as advanced engineering and environmental considerations. One approach for extracting the REEs from phosphate mining materials is by leaching with different extraction liquids. Previous studies have used concentrated strong acids to extract REEs from phosphate mining materials; however, the use of concentrated strong mineral acids comes with economic and environmental drawbacks, including long reaction times often at elevated temperatures [10–13]. To circumvent the disadvantages of strong mineral acids, weak organic acids can be used to produce REE complexes that improve dissolution [14]. In addition, organic acids can act as reducing agents and can control the oxidation state of the REEs, preventing the formation of insoluble REE oxides and leading to enhanced extraction concentrations [15].

REEs leached from phosphate mining materials need to be recovered from the extraction medium. Previous studies have used phosphoric acid-based solvent recovery chemicals or ion exchange resins to recover REEs from acid leach solutions [16,17]. Other investigations have attempted to improve this recovery process by removing non-REE impurities by magnetic separation or selective precipitation [18,19]. However, these methods have not yet seen widespread commercial acceptance because of various limitations, such as high cost, low efficiency, and the inability to economically extend the technology to large-scale operations.

Recently, the use of sorbents during the extraction process has been investigated. Rychkov and co-workers [4,20] reported the use of Purolite C160 as a cation exchange resin for the separation of REEs from phosphogypsum acidic suspensions. Purolite C160 is a polystyrenic, macroporous, strong acid cation resin that contains sulfonic acid functional groups. This resin has been shown to exhibit Freundlich isotherm behavior and a regeneration efficiency of more than 95% [21]. When used for the recovery of REEs from acid mine drainage, this resin had affinities for calcium and magnesium, and showed adsorption yields for the REEs of 85% to 89% [22].

In this study, we investigate the use of a chelating polymer, 1-octadecene, polymer with 2,5-furandione, sodium salt, to recover REEs from acid leach solutions. This chelation polymer is water-insoluble, and has been shown to possess novel metal adsorption characteristics [23]. It demonstrates metal adsorption capacities substantially higher than those of other heterogeneous adsorbents and almost equivalent to those obtained with homogeneous adsorbents. Additionally, the polymer exhibits pseudo-second-order kinetics, and has an adsorptive behavior accurately characterized by the Dubinin–Radushkevich isotherm model. The working pH range of the polymer is 1.5 to 14.

1-Octadecene, polymer with 2,5-furandione, sodium salt, contains dicarboxylic head groups that have high affinity for several polyvalent metal ions, including the transition metals and lanthanides, while showing low affinity for calcium and magnesium. In addition, the chelating polymer does not contain nitrogen, sulfur, or phosphorus; it can be efficiently filtered from aqueous solutions; and it can be regenerated for multiple recovery processes, making it environmentally friendly and economically attractive [3,23].

In the present study, we investigate the recovery of REEs from acidic extracts of four materials produced during phosphate mining: phosphate rock, phosphatic waste clay, phosphogypsum, and amine flotation tailings. The chelating polymer (1-octadecene, polymer with 2,5-furandione, sodium salt) is used to recover the REEs from acid extraction solutions.

2. Materials and Methods

Nitric acid (trace metal grade), citric acid anhydrous (lab grade), and ascorbic acid were purchased from Fisher Scientific (Waltham, MA, USA) and used as received without further purification. Phosphate rock, phosphogypsum, waste clay, and amine tailings were obtained from the Florida Institute of Phosphate Research (FIPR) Institute (Bartow, FL, USA). Poly(maleic anhydride-alt-1-octadecene) was purchased from Chevron Phillips Chemical Company (The Woodlands, TX, USA) and was converted to 1-octadecene, polymer with 2,5-furandione, sodium salt (also referred to as poly(2-octadecyl butanedioic acid)) [23]. The REE, uranium, and thorium content of the phosphate materials were determined by ACZ Laboratories, Inc. (Steamboat Springs, CO, USA) using EPA Methods M3050B and M6020. The REE, uranium, and thorium content of the extracts and polymer-treated filtrates were determined by inductively coupled plasma-optical emission spectroscopy (ICP-OES) using a Horiba Activa M ICP (Edison, NJ, USA). ICP standard solutions were purchased from Exatol Chemical Corporation (Clearwater, FL, USA). Moisture content was determined using a Mettler Toledo HB43-S Moisture Balance. Particle size was determined by a Microtrac S3500 laser diffraction particle analyzer (Microtrac, Inc., Montgomeryville, PA, USA).

2.1. Acid Extraction Methods

The extraction of REEs from the phosphate rock, phosphogypsum, waste clay, and amine tailings was performed in triplicate and carried out as follows: Each phosphate material was dried overnight in an oven at 100 °C. After drying, 40 g of a given material was mixed with 300 mL of the desired acid extraction solution (Table 1) under magnetic stirring for 15 min at room temperature (25 °C). After 15 min, the extraction solution was collected using vacuum filtration. The solid filter cake was then washed with 50 mL of the same acid extraction solution and washed again with 50 mL of deionized water. The filtrate was then collected and analyzed for REE, uranium, and thorium content using ICP-OES.

2.2. REE, Uranium, and Thorium Recovery Methods

The filtrates collected above were used to assess the REE metal recovery ability of 1-octadecene, polymer with 2,5-furandione, sodium salt. First, the pH of the filtrate was adjusted to pH 1.5 with sodium hydroxide. The pH values of the filtrates before adding NaOH were 1.44, 0.81, 0.89, and 0.79 for phosphate rock, phosphogypsum, waste clay, and amine tailings, respectively. Then, 5.0 mL of the pH-adjusted filtrate was diluted with 5.0 mL of deionized water, and 0.3 g (for phosphate rock, phosphogypsum, and waste clay) or 0.2 g (for amine tailings) of 1-octadecene, polymer with 2,5-furandione, sodium salt, was added to the diluted filtrate. The mixture was shaken for 1 h on a rotary shaker at 150 rpm and then syringe-filtered through 1.2 µm and 0.22 µm filters. The filtrates were analyzed for REE, uranium, and thorium content by ICP-OES.

2.3. Data Analysis

The REE, uranium, and thorium concentration in mg per g of material ($[REE,U,Th]_{mg/g}$) were calculated with the following equation:

$$[REE, U, Th]_{mg/g} = \frac{[REE, U, Th]_{ICP, ppm} \times Volume_L}{Mass\ of\ sample} \times Dilution\ factor, \quad (1)$$

where $[REE,U,Th]_{ICP,ppm}$ is the concentration from ICP-OES intensities in ppm and $Volume_L$ is the volume of acid used during the extraction in liters. A dilution factor was applied to the pH-adjusted extracts and the filtrates after polymer binding to account for the change in volume due to the addition of NaOH solution.

The standard deviation of the replicate extraction data was calculated using Equation (2):

$$S = \sqrt{(x_{i1} - x_{i2})^2 / 2}, \quad (2)$$

where x_{i1} and x_{i2} are the concentrations for REEs from separate individual extractions.

Table 1. List of Acid Extraction Solvents.

Acid Extraction Solution	Acid Composition
1	2.5% HNO_3
2	2.5% H_2SO_4
3	1.25% HNO_3 + 1.25% H_2SO_4 (2.5% total)
4	2.5% HNO_3 + 5% Citric Acid
5	2.5% HNO_3 + 5% Ascorbic Acid
6	2.5% H_2SO_4 + 5% Citric Acid
7	2.5% H_2SO_4 + 5% Citric Acid

3. Results and Discussion

3.1. Acid Extraction

The physical characteristics for the phosphate materials tested can be found in Table 2. The particle size of the waste clay was difficult to determine because of its high moisture content; then, after drying, it formed large particles several millimeters in size. The moisture content of the phosphate rock showed no loss on drying, because it was supplied having been previously dried by FIPR. The mineralogy of the samples tested have been reported [6,8,9,24–26].

Table 2. Physical Characteristics of Phosphate Rock and Phosphoric Acid Waste By-Products.

Sample Description	Sample Appearance	Average Moisture Content (before Drying)	Average Moisture Content (after Drying)	Average Particle Size (d50)
Phosphate Rock	Fine gray sand	1.14%	1.14%	257.3 μm
Phosphogypsum	Gray to beige powder	19.36%	0.89%	72.45 μm
Amine Tailings	Gray to brown wet powder	20.77%	0.23%	168.8 μm
Waste Clay	Gray	60.64%	1.70%	Not Determined

Figure 1 shows the amount of total REEs contained in each phosphate material. The levels of each individual REE can be found in Supporting Information. The highest levels of REEs were in phosphate rock, followed by waste clay, amine tailings, and phosphogypsum. The levels of REEs reported are commensurate with previous studies on Florida phosphate rock and phosphate wastes [1]. Phosphate rock contains the highest levels of REEs because it has the highest concentration of fluorapatite and the REE-bearing minerals monazite and xenotime.

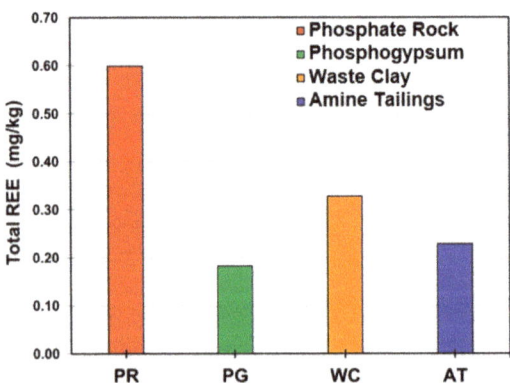

Figure 1. Total rare earth elements (REEs) contained in phosphate rock and phosphate waste materials.

One goal of this study was to identify potential extraction conditions that lead to high REE extraction efficiencies while being compatible with the extraction polymer, 1-octadecene, polymer with 2,5-furandione, sodium salt. High-temperature extractions are commonly advantageous for achieving higher extraction efficiencies; however, high temperatures can have adverse effects when dealing with phosphate rock and phosphate wastes. High-temperature acid extractions can lead to the increased dissolution of silicon and fluorine, which can lead to the formation of low-solubility REE fluorides [27]. Also at high temperatures, calcium sulfate will convert to the anhydrous state, which has a significantly lower solubility than the hydrated form [28]. Hence, room-temperature extractions were used throughout this study.

It should be noted that to assess the actual environmental and economic viability of these potential extraction methods, larger-scale studies would be required to address parameters such as waste moisture content (dewatering), particle size (milling), solid-to-liquid ratio, and mixing efficiency. This study was designed, in part, to determine the compatibility of the extraction polymer with previously reported acidic extract solutions of phosphate waste products rather than optimize the acid extraction process.

The results of the room-temperature extractions are shown in Figure 2. When using only mineral acids, the following trend was observed for the extraction efficiency of REEs: HNO_3 > HNO_3/H_2SO_4 > H_2SO_4. This trend was observed in all the phosphate materials tested. This can be attributed to the solubility of calcium sulfate, and is consistent with previous studies, including the recent work of Walawalker et al. [28]. Calcium nitrate has a higher solubility than calcium sulfate and is capable of dissolving more of the calcium minerals/salts; this leads to the better extraction of the REEs contained in the calcium minerals. While HNO_3 was more efficient at the extraction of the total REEs than H_2SO_4, some individual REEs showed an increased percentage of extraction when compared to the total REEs when using H_2SO_4, most notably, yttrium (Individual REE extraction data are contained in Supplementary Materials). When comparing the two REEs highest in concentration, cerium and yttrium, cerium showed an increased extraction in HNO_3 and an increased percentage when compared to the total REEs extracted (Figure 3a). Conversely, yttrium showed that a higher percentage was extracted in H_2SO_4 when compared to the other REEs (Figure 3b).

When adding organic acids, citric acid, when mixed with HNO_3, increased the extraction of REEs in only phosphate rock and phosphogypsum (Figure 2). When citric acid was used with H_2SO_4, it increased the REE extraction in phosphogypsum and amine tailings, but not phosphate rock or waste clay. Ascorbic acid showed an increased REE extraction in only waste clay, when using HNO_3. When using H_2SO_4, ascorbic acid had a slightly increased extraction in phosphate rock and phosphogypsum. Calcium ascorbate had a significantly higher solubility when compared to calcium citrate and calcium sulfate; hence, the increased solubility when adding ascorbic acid to H_2SO_4 can be attributed to the increased calcium dissolution from the phosphate materials, and is in agreement with the results of Mishelevich et al. [29]. However, as calcium has a high solubility in HNO_3, the addition of ascorbic acid had minimal effects on the REE extraction when added to HNO_3.

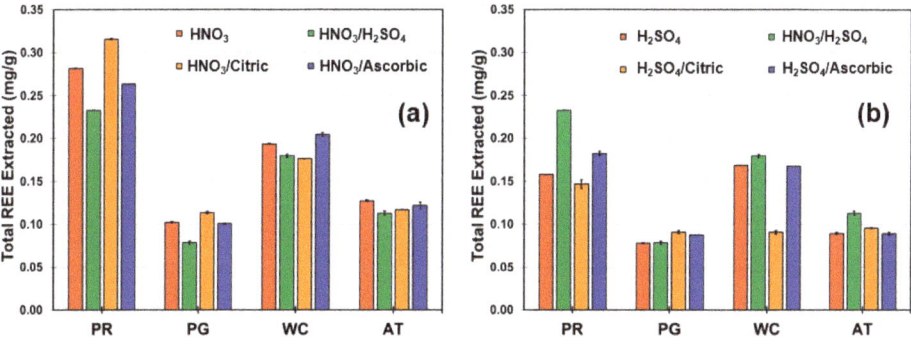

Figure 2. Extracted REEs using mixtures of (**a**) HNO_3, (**b**) H_2SO_4, citric acid, and ascorbic acid.

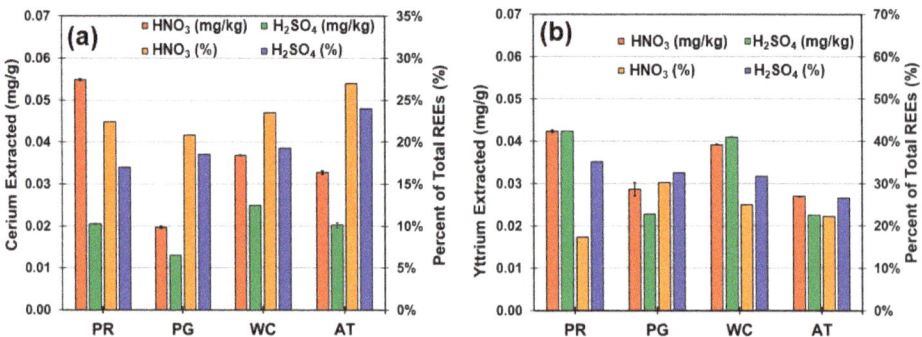

Figure 3. Extracted (**a**) cerium and (**b**) yttrium levels in HNO_3 and H_2SO_4.

Along with increasing the calcium/REE dissolution, ascorbic acid has the ability to act as a reducing agent; this ability could have an impact on cerium dissolution [15]. Cerium under mild aqueous conditions will occur in the Ce^{3+} state; however, at low pH and highly oxidative conditions, cerium can convert to the Ce^{4+} state, which forms insoluble Ce(IV) oxide [15]. In Figure 4, the addition of ascorbic acid to H_2SO_4 increased the cerium extraction in all phosphate materials; however, this could be a result of the increased calcium and total REE dissolution. When added to HNO_3, ascorbic acid only produced an increase in cerium extraction in waste clay (Figure 4). This could indicate more Ce(IV) being present in the waste clay when compared to the other phosphate materials.

Small amounts of uranium and thorium were co-extracted with the REEs in HNO_3. Extraction yields for uranium were 29%, 38%, 47%, and 16% from phosphogypsum, phosphate rock, waste clay, and amine tailings, respectively. Extraction yields for thorium were 26%, 0%, 38%, and 20% from phosphogypsum, phosphate rock, waste clay, and amine tailings, respectively (see Supplementary Materials).

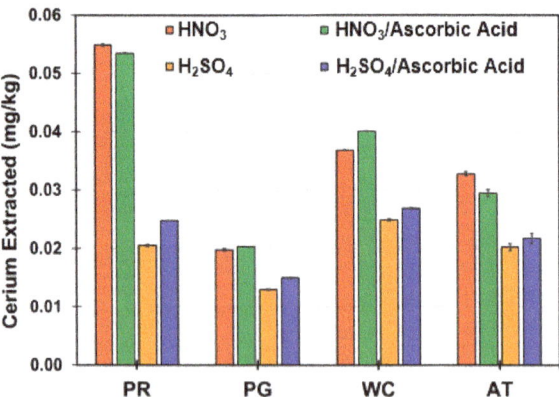

Figure 4. Extracted cerium levels with ascorbic acid.

3.2. REE, Uranium, and Thorium Recovery

REE recovery studies using 1-octadecene, polymer with 2,5 furandione, sodium salt, showed that the polymer bound REEs from all of the acid extract solutions. Due to the generally higher level of REEs extracted with the nitric acid solution, comprehensive REE recovery studies were performed using HNO_3 extraction solutions with the results being shown in Figure 5a (REEs present in high concentration) and Figure 5b (REEs present in low concentration).

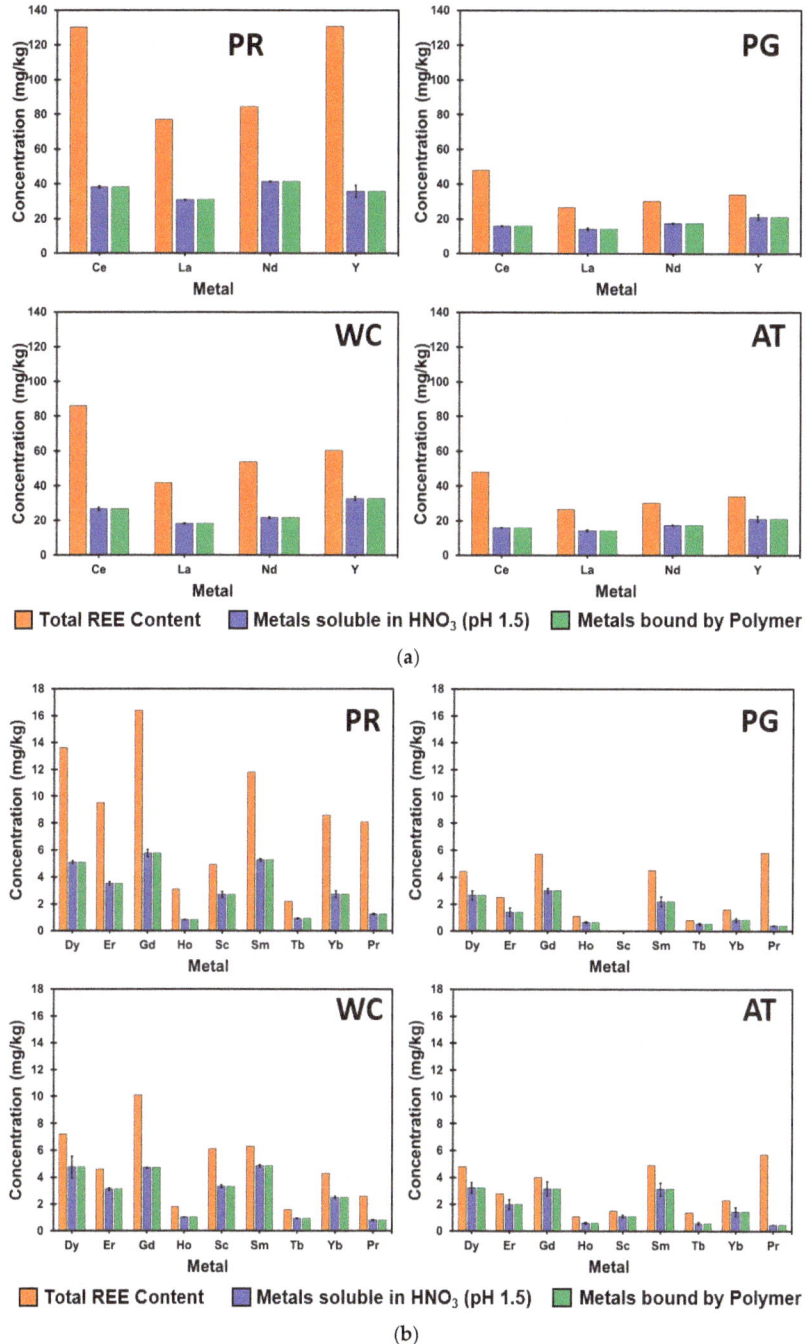

Figure 5. (**a**) High-concentration REE recovery using 1-octadecene 2,5-furandione, sodium salt, in phosphate rock (PR), phosphogypsum (PG), waste clay (WC), and amine tailings (AT). (**b**) Low-concentration REE recovery using 1-octadecene 2,5-furandione, sodium salt, in phosphate rock (PR), phosphogypsum (PG), waste clay (WC), and amine tailings (AT).

In this study, overall extraction and recovery yields for the rare earth elements were between 80% for gadolinium and 8% for praseodymium from amine tailings, between 70% for terbium and 7% for praseodymium from phosphogypsum, between 56% for scandium and 15% for praseodymium from phosphate rock, and between 77% for samarium and 31% for praseodymium from waste clay. Polymer recoveries for uranium were 23%, 31%, 47%, and 16% from phosphogypsum, phosphate rock, waste clay, and amine tailings, respectively. Polymer recoveries for thorium were 26%, 38%, and 20% from phosphogypsum, waste clay, and amine tailings, respectively. As noted in Section 3.1, thorium was not extracted by HNO_3 from phosphate rock under the conditions used in this study (see Supplementary Materials).

1-Octadecene, polymer with 2,5 furandione, sodium salt, bound 100% of the extracted REE in the acidic solution, representing a significant improvement to the yields of 85% to 89% previously reported for Purolite 160 [22]. This is likely due to the Freundlich isotherm behavior exhibited by Purolite 160 and other cation exchange resins, indicating monolayer coverage of the adsorbent surface by the metal, equal adsorbent affinity for all the binding sites, and that the adsorption at one site does not affect the adsorption at an adjacent site [21]. By contrast, 1-octadecene, polymer with 2,5 furandione, sodium salt, exhibits behavior best characterized by the Dubinin–Radushkevich isotherm with a mean free energy of adsorption of 31.6 $kJ \cdot mol^{-1}$, and a metal ion concentration-dependent Gibb's free energy of –6.50 to –11.61 $kJ \cdot mol^{-1}$. This indicates that the polymer has a heterogeneous sorbent surface with spontaneous chemisorption (major) and physisorption (minor) properties [23].

The extraction polymer was able to retain its ability to bind REEs in the presence of high levels of calcium and total dissolved solids. This is an improvement to the selectivity reported by Felipe et al. for Purolite 160 [22]. Additionally, the polymer was stable to corrosive, low-pH environments, including nitric acid solutions. As the extraction polymer is water-insoluble, this recovery process avoids the reported limitations of organic solvent contamination and emulsion formation associated with solvent extraction techniques for REE recovery [27]. Additionally, the insolubility of the extraction polymer in water avoids the high cost and technical limitations associated with organic solvent-soluble metal binding ligands [30].

Previous studies have shown that the chelating polymer can be recycled and used for multiple recovery processes [3,23]. We have successfully eluted the REEs from the polymer with dilute mineral acid solutions, and reused the polymer for subsequent recovery studies for 20 cycles without loss of function.

4. Conclusions

The two main objectives of this study were to a) identify potentially efficient REE extraction conditions that show compatibility with the extraction polymer, 1-octadecene, polymer with 2,5-furandione, sodium salt, and b) determine the ability of the extraction polymer to bind rare earth metals in a corrosive, low-pH environment containing high levels of calcium and other dissolved metals.

The extraction of REEs from phosphate rock and phosphate materials was investigated in dilute mineral acids with the addition of organic acids. In general, phosphate rock contained the highest levels of REEs and, under the extraction conditions employed in this study, HNO_3 was the most efficient acid for extracting REEs. Additional large-scale extraction and processing studies will need to be conducted to determine the optimal extraction conditions and perform accurate mass balance calculations. Based on the results of this and prior studies, numerous parameters, such as the moisture content of the waste solid, particle size of the solids, solid-to-liquid ratio, mode(s) of agitation, and means of separating the extraction solution from the waste solids will need to be optimized, along with determining if the extraction solution can be reused. As these and other parameters are likely to significantly impact both the performance and economic viability of this extraction process, an economic analysis of this step of the process was not possible in this study.

The REEs were recovered from HNO_3 extraction solutions using 1-octadecene, polymer with 2,5-furandione, sodium salt. At pH 1.5, the chelating polymer bound 100% of the REEs from phosphate rock, phosphogypsum, waste clay, and amine tailings, representing an improvement over strong acid cationic exchange resins such as Purolite 160.

The results of this study suggest that this polymer efficiently recovers rare earth elements from acidic extracts of phosphate mining waste products.

Supplementary Materials: The following are available online at http://www.mdpi.com/2075-163X/9/8/477/s1, Table S1: Polymer Binding and Percent Rare Earth Element Recovery.

Author Contributions: Conceptualization, J.P.L. and J.M.; Formal analysis, J.M. and Z.J.H.; Funding acquisition, J.P.L.; Investigation, J.M.; Methodology, J.P.L. and J.M.; Project administration, J.P.L.; Supervision, J.P.L.; Visualization, Z.J.H.; Writing—original draft, Z.J.H.; Writing—review & editing, J.P.L. and Z.J.H.

Funding: This work was funded in part by a grant from the Florida Industrial and Phosphate Research (FIPR) Institute, entitled FIPR #03-02-189R: Extraction and Recovery of Rare Earth Elements from Phosphate using Mx-107 and Chelok® Polymers.

Conflicts of Interest: The authors are current or former employees of Periodic Products, Inc. Intellectual property associated with 1-octadecene, polymer with 2,5-furandione, sodium salt is currently owned by Periodic Products, Inc.

References

1. Zhang, P. Comprehensive recovery and sustainable development of phosphate resources. *Procedia Eng.* **2014**, *83*, 37–51. [CrossRef]
2. Humphries, M. *Rare Earth Elements Global Supply Chain*; Diane Publishing: Collingdale, PA, USA, 2010.
3. Laurino, J.; Huba, Z.; Mustacato, J.; Zhang, P. The Recovery of Rare Earth Elements from Phosphate Rock and Phosphate Mining Waste Products Using a Novel Water-Insoluble Adsorption Polymer. In Proceedings of the Beneficiation of Phosphates VIII, Cape Town, South Africa, 29 April–4 May 2018.
4. Rychkov, V.N.; Kirillov, E.V.; Kirillov, S.V.; Semenishchev, V.S.; Bunkov, G.M.; Botalov, M.S.; Smyshlyaev, D.V.; Malyshev, A.S. Recovery of rare earth elements from phosphogypsum. *J. Clean. Prod.* **2018**, *196*, 674–681. [CrossRef]
5. Emsbo, P.; McLaughlin, P.I.; Breit, G.N.; du Bray, E.A.; Koenig, A.E. Rare earth elements in sedimentary phosphate deposits: Solution to the global REE crisis? *Gondwana Res.* **2015**, *27*, 776–785. [CrossRef]
6. Edahbi, M.; Benzaazoua, M.; Plante, B.; Doire, S.; Kormos, L. Mineralogical characterization using QEMSCAN® and leaching potential study of REE within silicate ores: A case study of the Matamec project, Quebec, Canada. *J. Geochem. Explor.* **2018**, *185*, 64–73. [CrossRef]
7. Soltani, F.; Abdollahy, M.; Petersen, J.; Ram, R.; Javad Koleini, S.M.; Moradkhani, D. Leaching and recovery of phosphate and rare earth elements from an iron-rich fluorapatite concentrate: Part II: Selective leaching of calcium and phosphate and acid baking of the residue. *Hydrometallurgy* **2019**, *184*, 29–38. [CrossRef]
8. May, A.; Sweeney, J.W. Assessment of environmental impacts associated with phosphogypsum in Florida. In *The Chemistry and Technology of Gypsum*; Kuntze, R.A., Ed.; ASTM International: West Conshohocken, PA, USA, 1984; pp. 116–139. ISBN 0-8031-0219-4.
9. Van Kauwenberg, S.J.; Cathcart, J.B.; McClellan, G.H. *Mineralogy and Alteration of the Phosphate Deposits of Florida*; U.S. Government Printing Office: Washington, DC, USA, 1914.
10. Weber, R.J.; Reisman, D.J. *Rare Earth Elements: A Review of Production, Processing, Recycling, and Associated Environmental Issues*; US EPA Region: Washington, DC, USA, 2012.
11. Zhang, P.; Liang, H.; Jin, Z.; DePaoli, D. The ultimate mineral processing challenge: Recovery of rare earths, phosphorus and uranium from Florida phosphatic clay. *Miner. Metall. Process.* **2017**, *34*, 183–188. [CrossRef]
12. Liang, H.; Zhang, P.; Jin, Z.; DePaoli, D. Rare-earth leaching from Florida phosphate rock in wet-process phosphoric acid production. *Miner. Metall. Process.* **2017**, *34*, 146–153. [CrossRef]
13. Canovas, C.R.; Chapron, S.; Arrachart, G.; Pellet-Rostaing, S. Leaching of rare earth elements (REEs) and impurities from phosphogypsum: A preliminary insight for further recovery of critical raw materials. *J. Clean. Prod.* **2019**, *219*, 225–235. [CrossRef]
14. Goyne, K.W.; Brantley, S.L.; Chorover, J. Rare earth element release from phosphate minerals in the presence of organic acids. *Chem. Geol.* **2010**, *278*, 1–14. [CrossRef]
15. Kolokolnikov, V.A.; Kovalev, M.I. Processing Rare-Earth Element Concentrate Obtained from Phosphogypsum. *Chem. Sustain. Dev.* **2009**, *17*, 261–266.

16. Wang, L.; Long, Z.; Huang, X.; Yu, Y.; Cui, D.; Zhang, G. Recovery of rare earths from wet-process phosphoric acid. *Hydrometallurgy* **2010**, *101*, 41–47. [CrossRef]
17. Al-Thyabat, S.; Zhang, P. REE extraction from phosphoric acid, phosphoric acid sludge, and phosphogypsum. *Miner. Process. Extr. Metall.* **2015**, *124*, 143–150. [CrossRef]
18. Battsengel, A.; Batnasan, A.; Haga, K.; Watanabe, Y.; Shibayama, A. Magnetic separation and leaching study of rare earth elements from apatite-iron ore. *Int. J. Soc. Mater. Eng. Resour.* **2018**, *23*, 88–92. [CrossRef]
19. Silva, R.G.; Morais, C.A.; Oliveira, E.D. Selective precipitation of rare earths from non-purified and purified sulfate liquors using sodium sulfate and disodium hydrogen phosphate. *Miner. Eng.* **2019**, *134*, 402–416. [CrossRef]
20. Rychkov, V.N.; Kirillov, E.V.; Kirillov, S.V.; Bunkov, G.M.; Mashkovtsev, M.A.; Botalov, M.S.; Volkovich, V.A.; Semenishchev, V.S. Selective ion exchange recovery of rare earth elements from uranium mining solutions. In *AIP Conference Proceedings*; American Institute of Physics: College Park, MD, USA, 2016.
21. Edebali, S.; Pehlivan, E. Evaluation of Cr(III) by ion-exchange resins from aqueous solution: Equilibrium, thermodymanics and kinetics. *Desalin. Water Treat.* **2014**, *52*, 7143–7153. [CrossRef]
22. Felipe, E.; Silva, G.; Vidigal, B.; Ladeira, A.C. Recovery of rare earth elements from acid mine drainage. In *Sustainable Industrial Processing Summit 2017 Volume 1: Barrios Intl. Symp/Non-ferrous Smelting & Hydro/Electrochemical Processing*; Flogen Star Outreach: Mont-Royal, QC, Canada, 2017; ISBN 978-1-987820-61-4.
23. Laurino, J.P. Removal of Lead (II) Ions by Poly (2-octadecyl butanedioic acid): Isothermal and Kinetic Studies. *J. Macromol. Sci. Part A Pure Appl. Chem.* **2008**, *45*, 612–619. [CrossRef]
24. Zhang, P.; Miller, J.; DePaoli, D.; Yang, M. Rare Earths Occurrence in Florida Phosphate Ore and Their Fate during Mining and Chemical Processing. In *Beneficiation of Phosphates VIII*; Zhang, P., Miller, J., Filho, L., Porteus, M., Snyders, N., Wingate, E., Akdogan, G., Eds.; ECI Symposium Series; Society for Mining, Metallurgy & Exploration: Englewood, CO, USA, 2018.
25. Zhang, P.; Bogan, M. Recovery of phosphate from Florida beneficiation slimes. *Miner. Eng.* **1995**, *8*, 523–534. [CrossRef]
26. Miller, J.; Lin, C.; Crossman, R. *Isolation and Characterization of Rare Earth Mineral Particles in Florida Phosphate Rock by de Rapid Scan Radiography and HRXMT*; Florida Industrial and Phosphate Research Institute: Bartow, FL, USA, 2015.
27. Preston, J.S.; Cole, P.M.; Craig, W.M.; Feather, A.M. The recovery of rare earth oxides from a phosphoric acid by-product. Part 1: Leaching of rare earth values and recovery of a mixed rare earth oxide by solvent extraction. *Hydrometallurgy* **1996**, *41*, 1–20. [CrossRef]
28. Walawalkar, M.; Nichol, C.K.; Azimi, G. Process investigation of the acid leaching of rare earth elements from phosphogypsum using HCl, HNO_3, and H_2SO_4. *Hydrometallurgy* **2016**, *166*, 195–204. [CrossRef]
29. Mishelevich, A.; Apelblat, A. Solubilities of magnesium-L-ascorbate, calcium-L-ascorbate, magnesium-L-glutamate, magnesium-D-gluconate, calcium-D-gluconate, calcium-D-heptagluconate, L-aspartic acid, and 3-nitrobenzoic acid in water. *J. Chem. Thermodyn.* **2008**, *40*, 897–900. [CrossRef]
30. Ellis, R.J.; Brigham, D.M.; Delmau, L.; Ivanov, A.S.; Williams, N.J.; Vo, M.N.; Reinhart, B.; Moyer, B.A.; Bryantsev, V.S. "Straining" to separate the rare earths: How the lanthanide contraction impacts chelation by diglycolamide ligands. *Inorg. Chem.* **2017**, *56*, 1152–1160. [CrossRef] [PubMed]

© 2019 by the authors. Licensee MDPI, Basel, Switzerland. This article is an open access article distributed under the terms and conditions of the Creative Commons Attribution (CC BY) license (http://creativecommons.org/licenses/by/4.0/).

Article

Phosphate Mine Tailing Recycling in Membrane Filter Manufacturing: Microstructure and Filtration Suitability

Mohamed Loutou [1,*], Wafa Misrar [2,3], Mohammed Koudad [1,4], Mohammed Mansori [2], Liga Grase [5], Claude Favotto [6], Yassine Taha [7] and Rachid Hakkou [2,7]

1. Département de Chimie, Faculté Pluridisciplinaire de Nador, Université Mohammed 1er, B.P. 300, Selouane, Nador 62700, Morocco; koudad.mohammed@yahoo.fr
2. LCME, Laboratoire de Chimie des Matériaux et d'Environnement, Faculté des Sciences et Techniques, Université Cadi Ayyad, Av. A. Khattabi, B.P. 549, Marrakech 40000, Morocco; misrar.wafa@gmail.com (W.M.); m.mansori@uca.ac.ma (M.M.); r.hakkou@uca.ma (R.H.)
3. LMCN, Laboratoire de Matière Condensée et Nanostructures, Faculté des Sciences et Techniques, Université Cadi Ayyad, B.P. 549, Marrakech 40000, Morocco
4. Laboratory of Applied Chemistry & Environment, Faculty of Sciences, Mohammed First University, Oujda 60000, Morocco
5. Institute of Silicate Materials, Faculty of Materials Science and Applied Chemistry, Riga Technical University, P. Valdena 3/7, LV-1048 Riga, Latvia; grase.liga@gmail.com
6. IM2NP, Institut Matériaux Microélectronique Nanosciences de Provence, Université de Toulon, Bât R, BP 20132–83957, La Garde-Cedex, France; claude.favotto@univ-tln.fr
7. Materials Science and Nano-engineering Department, Mohammed VI Polytechnic University (UM6P), Lot 660-Hay Moulay Rachid, Bengurir, Morocco; yassine.taha@um6p.ma
* Correspondence: mohamedloutou@yahoo.fr; Tel.: +212-672-057-790

Received: 20 February 2019; Accepted: 21 May 2019; Published: 24 May 2019

Abstract: Ceramic membrane filters based on industrial by-products can be considered to be a valorization alternative of phosphate mine tailings, even more so if these ceramic membranes are used in the industrial wastewater treatment due to their good mechanical, chemical, and thermal resistance. The depollution of textile industry rejections with this method has not been studied in detail previously. In this work, ceramic membrane filters have been manufactured from natural clay and phosphate mine tailings (phosphate sludge). Blends of the abovementioned materials with a pore-forming agent (sawdust, up to 20 wt. %) were investigated in the range 900–1100 °C using thermal analysis, X-ray diffraction, scanning electron microscopy, and mercury porosimetry. Ceramic properties were measured as a function of firing temperature and sawdust addition. Filtration tests were carried out on samples with advantageous properties. The results showed that gehlenite together with diopside neoformed from lime decomposed carbonates and breakdown products of clay minerals, while calcium phosphate derived from partial decomposition of fluorapatite. Both quartz and fluorapatite resisted heating. The results of the experimental design showed that the variations of physical properties versus processing factors were well described by the polynomial model. Filtration results are quite interesting, allowing these membranes to be used in industrial effluent treatment.

Keywords: ceramic membrane; phosphate mine tailings; industrial waste; filtration; experimental design

1. Introduction

In recent years, ceramic filtering membranes have been used in a wide range of applications and processes such as biotechnology, pharmaceuticals, the food industry, and industrial effluent treatment [1–4]. The growing interest that membrane filters has received is linked to their thermal, chemical, and mechanical properties. These remarkable properties offer ceramic membranes a large advantage over their polymeric [5,6] and metallic counterparts. However, they are usually more expensive than those based on polymers. In this respect, research on new ceramic materials that are cheaper for membrane filter manufacturing should benefit from extensive use of membrane technology, especially in developing countries, where many environmental issues should be addressed at low cost.

Micaceous clay and phosphate sludge obtained from clay deposits and phosphate-discharge plant dams (Morocco) respectively [7,8] could be a suitable material for low-cost membrane manufacture [9,10]. In fact, this method presents the advantage of allowing the substitution of the materials commonly used in this field (alumina, zirconia, cordierite, mullite etc.) [11–15] by other, less expensive ones (clay) that have similar properties. Also, there is the fact that it is an excellent way of managing industrial waste, which may constitute a potential source of pollution [8]. It has been logged that significant amounts of the abovementioned waste can be used in some specific and/or common applications, such as lightweight aggregates [16], soil amendments [17], or even construction [18].

Several scientific works have been performed regarding the incorporation of aluminosilicate-based materials in the filtration membrane manufacture [19–24]. The nature of the raw material though affects both physical and chemical properties of the final ceramic product. These properties can be tailored for each particular application by controlling the chemical and mineralogical compositions [25], and the microstructure of the used materials and additives. Despite this, little attention has been given to the use of natural pore-forming agents (lime [26], starch [27,28], wood [29], organic waste (paper from the paper industry) [30], sawdust from the woodworking industry [31–33], etc.).

This study is devoted to investigating the feasibility of the manufacture of new ceramic membrane filters from natural micaceous clay and industrial by-product (phosphate sludge). These membranes are designed to be used in industrial effluent treatment.

2. Materials and Experimental Techniques

2.1. Materials

The starting raw materials used in this research were a naturally occurring clay (SA) and phosphate waste (PS). The micaceous clay was extracted from a clay stratum in the region of Safi (Morocco) known by the pottery industry. Phosphate waste was obtained from phosphate sludge ponds generated from beneficiation plants of Moroccan phosphate rocks (Youssoufia, Morocco). The mineralogical compositions of the abovementioned materials are provided in Table 1 and Figure 1, respectively. SA consisted of hydro-muscovite (25 wt. %), quartz (12 wt. %), and dolomite (63 wt. %). PS was composed of quartz (17 wt. %), fluorapatite (44 wt. %), calcite (15 wt. %), dolomite (7 wt. %), and a smectite clay mineral (7 wt. %).

Cedar sawdust (SC) was used in this study as a pore-forming agent. It was supplied by a local carpentry factory (Marrakech, Morocco). All the raw materials used were sieved through a gyratory sieve (100 µm).

Table 1. Mineralogical compositions (wt. %) of studied raw materials. SA: Micaceous raw clay and PS: phosphate sludge. F: Fluorapatite, Q: Quartz, C: Calcite, M: Muscovite, D: Dolomite.

	Mineralogical Composition				
	F	Q	D	C	M
PS	44	17	7	15	-
SA	-	12	63	-	25

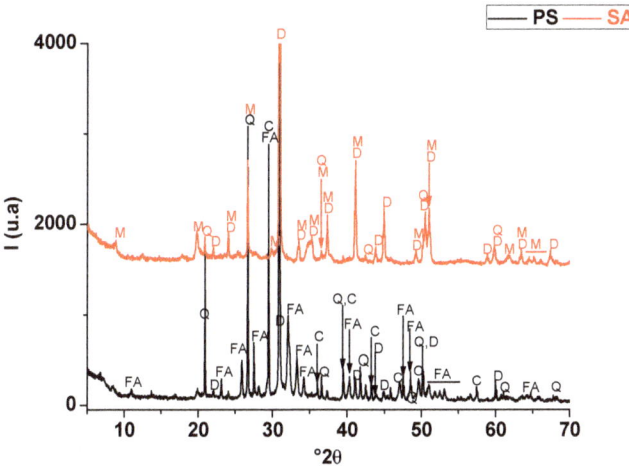

Figure 1. X-ray diffraction patterns of phosphate sludge (PS) and a dolomite-rich clay (SA). FA: fluorapatite (PDF #71-0880); Q: quartz (PDF # 5-0490); D: dolomite (PDF # 83-1766); C: calcite (PDF # 72-1650); M: Muscovite (PDF # 43-0685).

2.2. Experimental Techniques

Two binary mixtures (SA-SC (up to 20 wt. % of SC) and PS-SC (up to 20 wt. % of SC)) and a ternary one (clay-phosphate sludge-sawdust (SA:PS = 50:50 and up to 20 wt. % of SC)) were prepared for the present study. The materials were dry blended in a mortar before being moistened with tap water (10 wt. % moisture) to have consistent and comparable specimen. The mixtures were then shaped into cylindrical form (D = 40 mm and h = 3 mm) for the filtration tests and prismatic (L = 60 mm, W = 10 mm and h = 5 mm) for the mechanical properties, and this using an uniaxial laboratory-type pressing in a suitable mold (compression pressure = 2 tons). As reference, pore-forming agent-free samples have been prepared. The samples were heated starting from room temperature in an electric furnace (Nabertherm) at a rate of 5 °C/min in the range 900–1100 °C for 4 h and then cooled to room temperature in the switched-off furnace.

Samples of the heated blends were analyzed with X-ray diffraction (XRD) using an Empyrean PANalytical diffractometer operating with copper radiation (Kα(Cu) = 1.5418 Å). Quantitative mineralogical analysis was performed using the RIR method (reference intensity ratio). The thermal analysis was performed by a Setaram Setsys 24 apparatus (atmosphere: air; heating rate: 10 °C/min, reference material: Alumina). The morphological features of the blends were studied on fracture surfaces sputtered with Cr by a Schottky field emission scanning electron microscope (FE-SEM) (Nova NanoSEM 650, FEI Company, Eindhoven, The Netherlands) coupled with an energy dispersive spectroscopy (EDS) (TEAM ™ integrated EDS with an Apollo X silicon drift detector) for quantitative X-ray microanalysis. Mercury intrusion porosimetry (Pore Master 33, Quantachrome Instruments) was used to determine the pore volume distribution. It is established on the basis that a non-wetting liquid (any with a contact angle superior than 90°) will only intrude into capillaries under pressure. The relationship between the pressure and capillary diameter is described by Washburn [34] as Equation (1):

$$P = \frac{-4\,\gamma\,\cos(\theta)}{d} \quad (1)$$

where P: pressure, γ surface tension of the liquid, θ: contact angle of the liquid, and d: diameter of the capillary. Mercury must be forced using pressure into the pores of a material. The pore-size distribution is determined from the volume intruded at each pressure increment. Total porosity is determined from the total volume intruded.

The bending strength was performed with an Instron 3369 apparatus. The load and loading used speed were 50 kN and 0.1 mm/min, respectively. For this trial, five heated samples were studied.

Filtration tests were conducted on a laboratory microfiltration pilot, using a recycling configuration. The pilot was equipped with two silver wire electrodes used to measure the diffusion potential. The diffusion potential coefficient is defined by the Equation (2) [35]:

$$DP = \frac{\Delta E}{\Delta P} \quad (2)$$

where ΔE: the electric potential between the walls of the membrane and ΔP the applied pressure. The tests were performed at room temperature. The membranes were conditioned by immersion in distilled water for 12 h before the filtration tests. The schematic diagram of the filtration pilot is shown in Figure 2. It was principally composed of a circulation pump, air compressor, feed container of two liters, two manometers, and a membrane model. Transmembrane pressure was variable via a pressure regulator. The filtering surface area was about 24 cm² for all filtration samples. It is worth noting that three membrane samples were employed for flirtation tests to obtain the reproducibility of experimental results. Moreover, all filtration experimentations were conducted at room temperature.

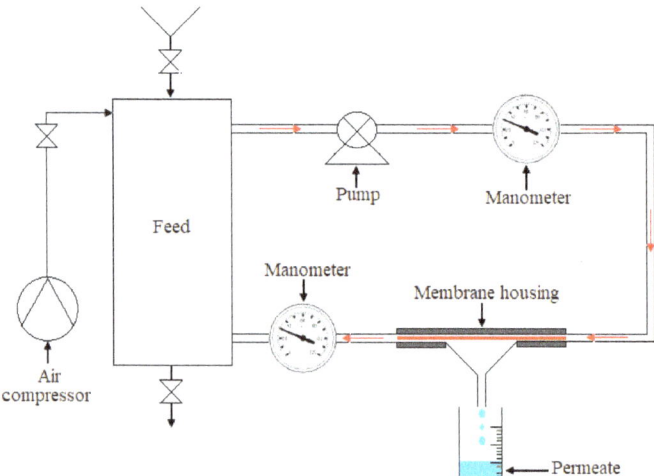

Figure 2. Scheme of the filtration pilot.

The chemical oxygen demand (COD) was determined using a LOVIBOND PCcheckit vario (LOVIBOND, London, UK), which contains a photometer and an ET 108 reactor. 2 mL of samples was mixed with the oxidizing acid solution in a vial that was then held at 150 °C for 2 h. After cooling, the mixed solution was analyzed in the PCcheckit vario photometer.

The biological oxygen demand (BOD) was measured using a LOVIBOND IR-Sensomat containing an IR-pressure-sensor and a stirring system. 500 mL of every sample was kept in a flask in an incubator for 5 days. Variance in air pressure was detected by the IR-sensor and converted directly into mg/L of BOD.

Suspended solids content was measured using a DR2010 portable data logging spectrophotometer (photometric method).

2.3. Experimental Design

The variations of the technological properties (Y_i) of the membranes versus the processing factors (sawdust addition (τ), temperature (T), and soaking time (t)) was assessed using a second-degree polynomial model [16,36,37]. This equation is in the following form:

$$Y = a_0 + a_1X_1 + a_2X_2 + a_3X_3 + a_{11}X_1^2 + a_{22}X_2^2 + a_{33}X_3^2 + a_{12}X_1X_2 + a_{13}X_1X_3 + a_{23}X_2X_3 \quad (3)$$

$$Y = a_0 + \sum_{i=1}^{k} a_i X_i + \sum_{i=1}^{k} a_{ii} X_i^2 + \sum_{i=1}^{k-1} \sum_{j=i+1}^{k} a_{ij} X_i X_j$$

where: X_1, X_2 and X_3 are the coded variables corresponding to τ, T and t, respectively.

$$X_1 = \frac{\tau - \tau_0}{\Delta \tau}; \quad X_2 = \frac{T - T_0}{\Delta T}; \quad X_3 = \frac{t - t_0}{\Delta t} \quad (4)$$

τ_0, T_0, and t_0 are the sawdust addition, firing temperature, and soaking time at the centers of the experimental range ($\tau_0 = 12.5$ wt. %, $T_0 = 1000$ °C and $t_0 = 2.5$ h). Δt, ΔT, and Δt are the variation steps of the considered variables ($\Delta t = 7.5$ wt. %, $\Delta T = 100$ °C and $\Delta t = 1.5$ h). a_0 is a constant, and a_1, a_2, and a_3 are the weights of the effects of sawdust addition, temperature, and soaking time, respectively. a_{ij} expresses the weight of the interaction effect between i and j factors, and a_{ii} is considered to be a curve-shaped parameter. The coefficients were calculated by Nemrod software using the least-squares regression [38,39]. For this purpose, multiple experiments (16) were performed according to the Dohlert matrix. The test at the center was multiplied (repeated 3 times) to estimate the experimental error. The experiments proposed by Doehlert matrix and the experimental values of physical properties are shown in Table 2. The accuracy and validity of the used model was confirmed by the analysis of variance (ANOVA) [40–42] represented in Table 3.

Table 2. Experimental design matrix (Doehlert) and measured values of the studied properties. Y_1: density; Y_2: firing shrinkage; Y_3: water absorption; Y_4: bending strength.

X_1	X_2	X_3	τ (wt. %)	T (°C)	t (h)	Density Y_1 (g/cm³)	Firing Shrinkage Y_2 (%)	Water Absorption Y_3 (%)	Bending Strength Y_4 (MPa)
1	0	0	20.00	1000.00	2.50	2.012	2.010	22.763	1.218
−1	0	0	5.00	1000.00	2.50	2.050	1.050	16.792	1.506
0.5	0.866	0	16.25	1086.60	2.50	2.018	2.677	22.218	1.760
−1	−0.866	0	8.75	913.40	2.50	2.030	0.676	20.429	1.610
0.5	−0.866	0	16.25	913.40	2.50	2.065	2.103	24.497	1.210
−1	0.866	0	8.75	1086.60	2.50	2.100	1.643	18.603	1.970
0.5	0.2887	0.8165	16.25	1028.87	3.72	2.010	2.309	22.938	1.590
−1	−0.289	−0.817	8.75	971.13	1.28	2.077	0.773	18.878	1.680
0.5	−0.289	−0.817	16.25	971.13	1.28	2.140	2.199	23.360	1.480
0	0.5774	−0.817	12.50	1057.74	1.28	2.065	1.633	20.217	1.580
−1	0.2887	0.8165	8.75	1028.87	3.72	2.087	0.883	18.644	1.800
0	−0.577	0.8165	12.50	942.26	3.72	2.040	0.961	22.511	1.300
0	0	0	12.50	1000.00	2.50	2.040	1.537	22.058	1.460
0	0	0	12.50	1000.00	2.50	2.100	1.500	21.980	1.400
0	0	0	12.50	1000.00	2.50	1.990	1.560	22.100	1.470
0	0	0	12.50	1000.00	2.50	2.038	1.56	22.1	1.47

Table 3. Of variance (ANOVA), values of correlation coefficient (R^2) and standard deviations (σ) for the considered responses: Y_1: density; Y_2: firing shrinkage; Y_3: water absorption; Y_4: bending strength.

	Water Absorption			
	Y_1 (g/cm^3)	Y_2 (%)	Y_3 (%)	Y_4 (MPa)
F-ratio	70.5559	1512.0456	86.5970	350.329
Signification	0.164 **	<0.01 ***	0.205 **	0.0352 ***
R^2	0.98	0.99	0.97	0.96
σ	0.1	0.084	0.15	0.12

*** Statistically significant at the level > 99.9% (probability (p) < 0.001). ** Statistically significant at the level > 99% (p < 0.01).

3. Results and Discussion

3.1. Thermal Transformations and Microstructure of Membranes

3.1.1. Micaceous Clay-Sawdust Mixture (SA-SC)

The X-ray diffraction pattern of the heated materials (Figure 3) showed that hydro-muscovite and dolomite decomposed at T < 900 °C. Indeed, hydro-muscovite sheet mica dehydroxylated at T < 700 °C and dolomite decomposed in the range 750–880 °C. The Differential thermal analysis (DTA) analysis (Figure 4) supports this results and indicates the occurrence of tow peaks at 787 °C and 878 °C corresponding to the decomposition of dolomite in two stages: $(CaMg(CO_3)_2 \rightarrow CaCO_3 + CO_2 + MgO$ and $CaCO_3 \rightarrow CaO + CO_2)$ [43,44]. Quartz resisted heat treatment, but its amount diminished slightly with increasing temperature, likely because it contributed to the neoformation process. Moreover, gehlenite and diopside were detected at 900 °C, likely from the breakdowns of clay mineral (hydro-muscovite) and released lime of carbonate (dolomite) decomposition. As far as the X-ray diffraction was concerned, the amount of gehlenite decreased, and that of diopside increased with increasing temperature, suggesting that the latter developed with further heat treatment. The added pore-forming agent (sawdust) seems not to have influenced the neoformation process in either qualitative or quantitative terms.

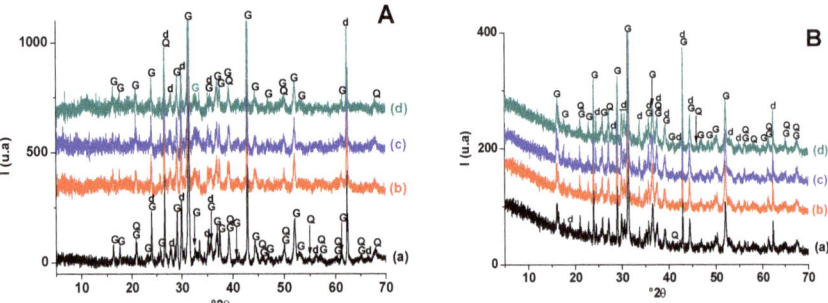

Figure 3. X-ray diffraction patterns of blends heated at 900 °C (**A**) and 1100 °C (**B**). (a) SA; (b) SA-SC (5 wt. % of SC); (c) SA-SC (10 wt. % of SC); (d) SA-SC (20 wt. % of SC). G: Gehlenite (PDF # 72-2128); Q: Quartz (PDF # 79-1911); d: diopside (PDF # 75-0945).

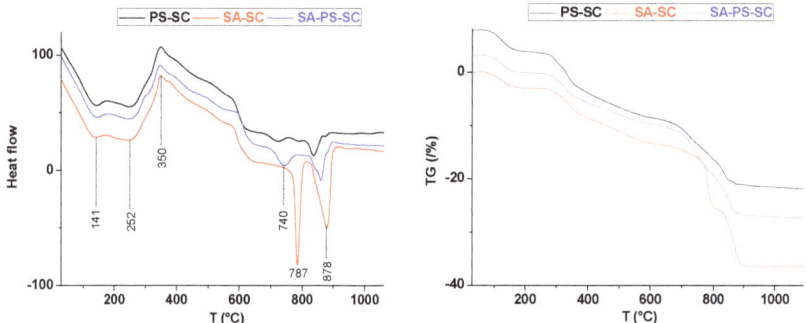

Figure 4. Thermal curves (DTA) and thermogravimetric analysis (TG) of PS-SC, SA-SC and SA-PS-SC.

As a result of heating, the laminated structure detected at low temperature (Figure 5a) disappeared, and agglomerated particles took place at high temperature (Figure 5b). The SEM observations (Figure 5c,d) also revealed grainy particles across the matrix, which corresponded to grains of gehlenite according to Energy Dispersive Spectroscopy (EDS) analysis. Loutou et al. (2013) [8] reported that gehlenite formed as granular particles within pores. These pores appeared to correspond to carbonates sites rich with lime (gehlenite formation source). At high temperature (1100 °C), grainy particles of gehlenite were agglomerated (Figure 5e,f) leading to the coalescence of some pores with the increase of temperature.

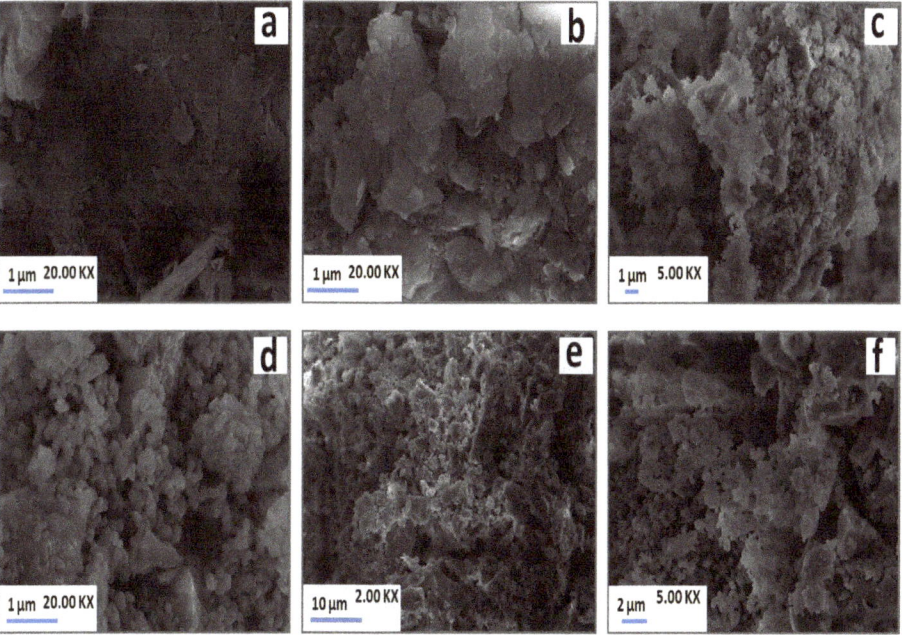

Figure 5. SEM micrographs of PS-SC (20 wt. % of SC) blends heated at 900 °C (**a**) and 1100 °C (**b**–**f**).

3.1.2. Phosphate Sludge-Sawdust Blend (PS-SC)

Referring once again to Figure 4, the thermal analysis (DTA) of PS-SC-fired blends showed that samples experienced four transformations at 141, 252, 730, and 836 °C. These endothermic peaks were attributed respectively to hygroscopic water loss [45], pore additive (sawdust) dehydration, dolomite first stage decomposition, and calcite decomposition [46,47]. The addition of the sawdust shifted these peaks to a higher temperature. The exothermic peak located at 350 °C was ascribed to sawdust firing, and was accompanied with 1.61% of weight loss (Figure 4).

Taking into consideration the X-ray diffractograms (Figure 6), only quartz and fluorapatite resisted the heat treatment. Their amount decreased with increasing temperature, probably due to the involvement of quartz in the neoformation and the partial fusion of fluorapatite at high temperatures (1100 °C). Gehlenite was the only phase neoformed in all studied PS-SC samples, and its proportion increased at the expense of both quartz and fluorapatite.

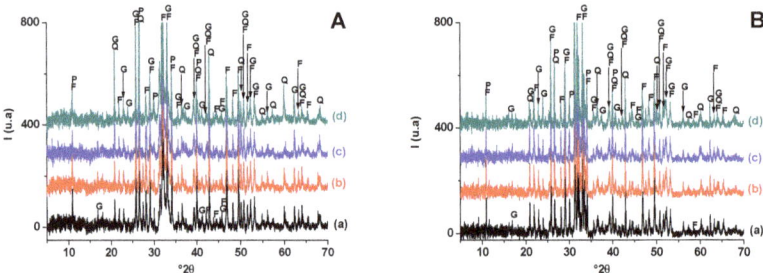

Figure 6. X-ray diffraction patterns of SA-PS (50 wt. % of SA) blends heated at 900 °C (**A**) and 1100 °C (**B**). (a) SA-PS; (b) SA-PS-SC (5wt. % of SC); (c) SA-PS-SC (10 wt. % of SC); (d) SA-PS-SC (20 wt. % of SC). G: Gehlenite (PDF # 72-2128); Q: Quartz (PDF # 79-1911); F: Fluorapatite (PDF # 83-0556); P: Calcium phosphates (PDF # 11-0232).

SEM examinations showed certain porosity with isolated particles at low temperature (Figure 7a). Apparently, the latter particles coalesced with the increase of temperature leading to coarse melted blocks throughout the samples (Figure 7b).

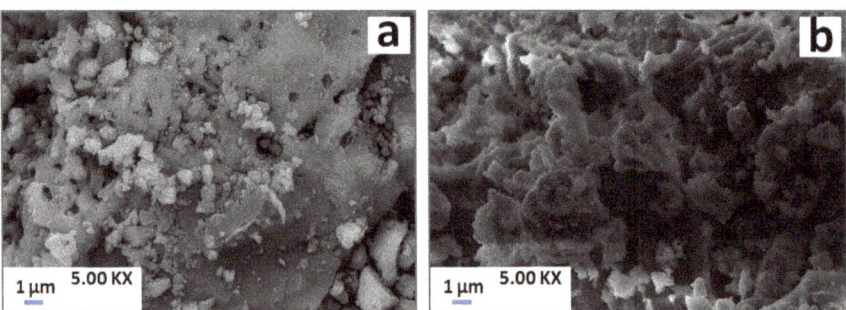

Figure 7. SEM micrographs of PS-SC (20 wt. % of SC) blends heated at 900 °C (**a**) and 1100 °C (**b**).

3.1.3. Ternary Mixture (PS-SA-SC)

X-ray diffraction analysis of fired samples (Figure 8.) showed that: (i) quartz and fluorapatite (original minerals) were encountered in all heated samples, but their amounts varied with firing temperature; (ii) hydro-muscovite and carbonate decomposed at T < 900 °C; (iii) the addition of the sawdust did not affect phase transformation; (vi) gehlenite and calcium phosphate developed at lower temperature and resisted heating afterwards. Gehlenite was derived from carbonate-released lime and the breakdowns of clay minerals, while calcium phosphate resulted from fluorapatite partial decomposition.

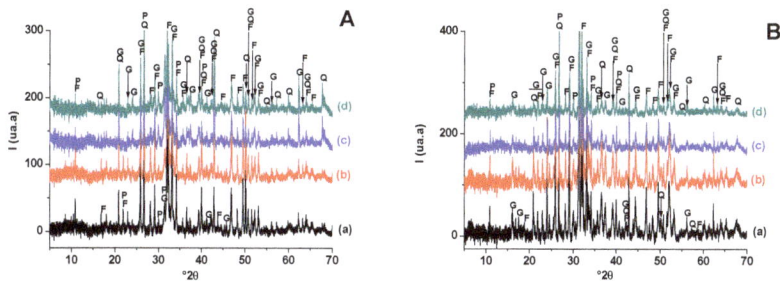

Figure 8. X-ray diffraction patterns of SA-PS (50 wt. % of SA) blends heated at 900 °C (**A**) and 1100 °C (**B**). (a) SA-PS; (b) SA-PS-SC (5 wt. % of SC); (c) SA-PS-SC (10 wt. % of SC); (d) SA-PS-SC (20 wt. % of SC). G: Gehlenite (PDF # 72-2128); Q: Quartz (PDF # 79-1911); F: Fluorapatite (PDF # 83-0556); P: Calcium phosphates (PDF # 11-0232).

According to the recorded Thermogravimetric analysis (TGA) curves (Figure 4), two weight losses occurred of about 12.48% and 7.77% between 700 and 880 °C, respectively. In fact, these weight losses consisted of two distinct stages. The first is primary decomposition of dolomite, whereas the second is related to the CO_2 release as a result of calcite decomposition. The weight-loss ratio of the last stage is more prominent.

Due to the heat treatment, grainy structures detected by SEM (Figure 9a) underwent partial melting and thereby contributed to melt formation (Figure 9c). On the other hand, those of gehlenite were agglomerated and was well crystallized (Figure 9b).

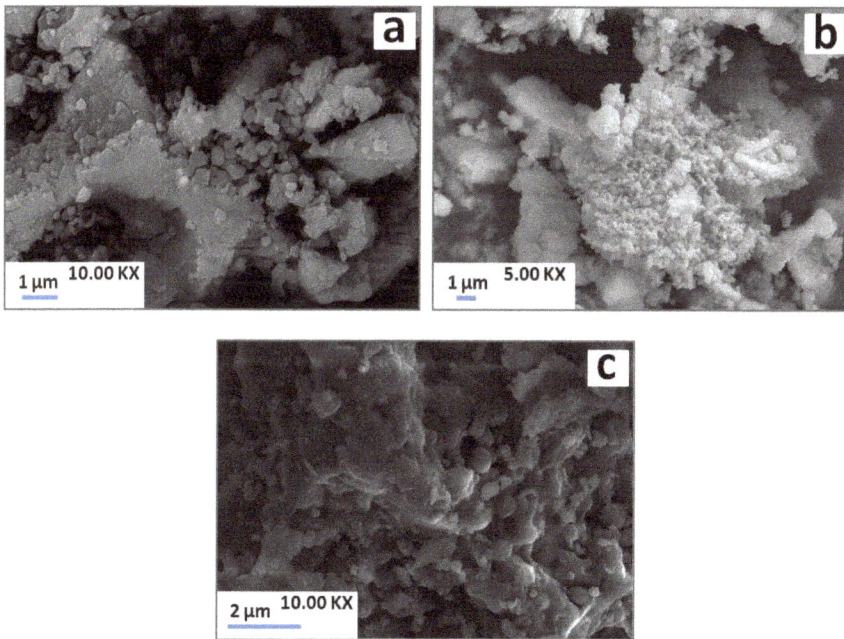

Figure 9. SEM micrographs of SA-PS-SC blends heated at 900 °C (**a,b**) and 1100 °C (**c**).

3.2. Porosimetry and Filtration Tests

Membranes prepared at 900 °C were used for filtration tests to keep a certain porosity level and avoid the partial melting of certain constituents at high temperatures.

Results of pore-size distribution (mercury-intruded volume versus pore size), for micaceous clay, phosphate sludge, and the ternary mixtures, are shown in Figure 10. It is revealed that both PS-containing samples (PS-SA and PS-SA-SC) displayed both larger pores (12–7 μm diameter) and lower pores (2–0.2 μm diameter) with a narrow range of distribution, and therefore exposed a marked incremental porosity (50% and 38%) for pores with 10 μm diameter respectively (Figure 10b,c). On the other hand, Figure 10a showed that SA-SC samples were the subject of only micropore formations. An incremental porosity of about 23% corresponding to pores of 0.7 μm diameter. The results obtained for both PS-containing blends is very interesting, since it will help inhibit the membrane resistance against mass transfer, and subsequently increase the filtration performance. It is clear that the pore formation process depends closely on the firing temperature [48,49] and elements liable to form pores (organic additives, carbonates, etc.).

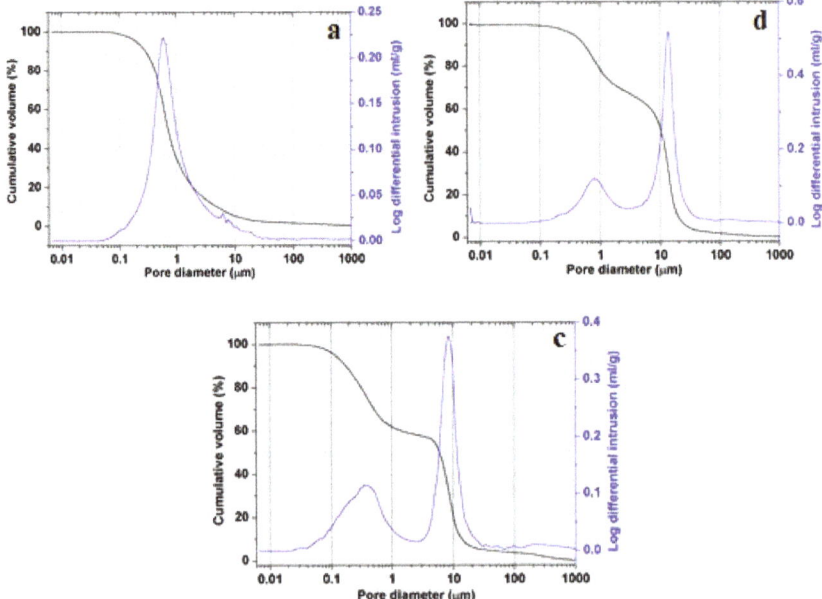

Figure 10. Pore-size distribution from mercury porosimetry of SA-SC blends (**a**), PS-SC blends (**b**) and ternary mixture (SA-PS-SC) (**c**) heated at 900 °C.

The permeability-measuring results from mercury porosimetry are reported in Table 4. Permeability coefficient values showed a marked difference between the behavior of the clay and the phosphate sludge. The coefficients corresponding to the sludge are twice those of clay (K_{PS}/K_{SA} = 2). This is probably because the sludge contains more carbonate content (element contributing to pore formation) leading to the occurrence of an additional amount of porosity. These observations also support the role of pore size and the emergence of a reduced tortuous path and further permeability to the water flow in the membrane. The pore morphology also has an effect in this respect, and is to be taken into account.

Table 4. Permeability results from mercury porosimetry (with and/or without tortuosity effect).

Coefficient of Permeability (nm^2)	SA-SC	PS-SC	SA-PS-SC
K_n (Neglecting Tortuosity Effects)	1.2040×10^{-01}	2.0526×10^{-01}	1.3857×10^{-01}
K_t (With Tortuosity Effects)	1.1485×10^{-01}	2.0158×10^{-01}	1.3597×10^{-01}

Variations of the flux versus the transmembrane pressure (Figure 11) showed a nearly linear behavior for all the studied samples. Theoretically, the flux is defined by the following equation:

$$J = L_P \times \Delta P + Cste$$

where J is the permeate flux (L/h.m^2), L_p water permeability (L/h.m^2.bar), and ΔP the applied transmembrane pressure (bar). The value of the water permeability can be deduced from the slope of the line drawn through the experimental points. It is about 93.89, 68.89, and 52.55 L/h.m^2.bar for SA-SC, PS-SC, and SA-PS-SC, respectively.

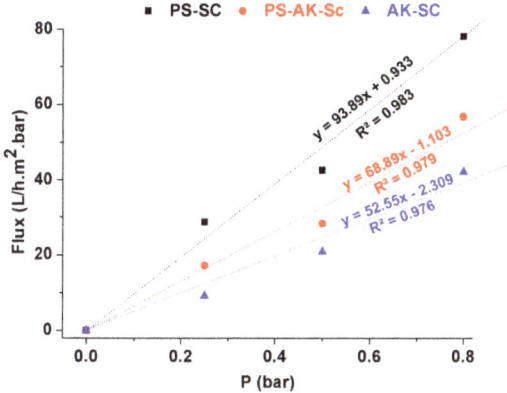

Figure 11. Variation of water flux as a function of transmembrane pressure.

Results of tangential filtration test for textile effluent are represented in Figure 12. It showed the variations of textile effluent permeate with filtration time at a pressure of 0.25 bar. The permeate flux declined continuously during the filtration test for all membranes. This reduction in flow is due essentially to the accumulation of suspended particles onto the membrane surface.

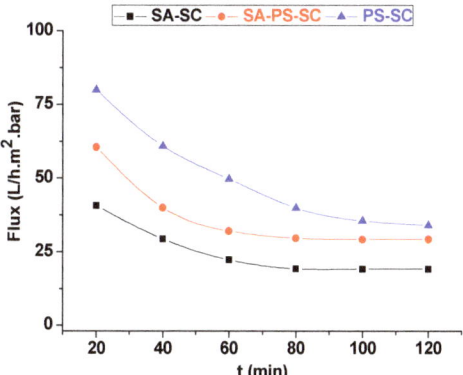

Figure 12. Permeate of textile effluent versus filtration time.

The characterization results of the wastewater samples are reported in mean values (min, max) in Table 4. Filtration suitability was also investigated by assessing the COD present in the industrial wastewater. Generally, the effects resulting from concentration polarization can be minimized but not canceled. In fact, during the movement of the wastewater through the membrane during filtration, particles of different sizes settle on the surface of the membrane. This leads to the formation of a gel-like pseudo-layer. The latter has the effect of reducing the permeate flux due to the increase in hydraulic resistance of the system (as explained before). Depending on the size and/or the structure of the pores, as well as their surface distribution, a total blockage of the flux can occur. Simultaneously with the preceding phenomena, the pseudo-layer often leads to a gradual increase in solutes retention, as can be seen in COD values before and after filtration in Table 4. According to Lopes et al. [50], the performance of a membrane is satisfactory when the COD reduction exceeds 73%. In our case, the COD reduction was in the range 70, 72, and 75%, which corresponds to a COD of 405, 380, and 340 mg/L, respectively. The ternary mixture-based membrane was the only one that fell within the acceptance range. The COD remaining in the filtrate probably came from low-molecular-weight solutes that may have passed through the membrane. It should be noted that COD retention is highly influenced by temperature, transmembrane pressure, and pollutant concentration [50–53].

Similarly, the removal efficiency of total organic carbon (TOC) and BOD were in the range 77–83% and 89–93%, respectively for the studied blends. Of all mixtures, the ternary one exhibited the best measurements.

3.3. Physical Properties and Effect of Processing Factors

Given the limitations to the number of figures, only the SA-PS-SC mixture heated at different temperatures will be treated.

As explained in the experimental procedure section, a Dohlert matrix was used for creating the experimental design. The experimental conditions of the planned experiments and the measured properties are given in Table 5.

Table 5. Chemical analyzes of wastewater before and after filtration and comparison with literature values.

Chemical analyzes	Previous Studies (Literature)		Before Filtration	Current Study		
				After Filtration		
	Textile Wastewater	Dyeing Wastewater		PS-SC	SA-SC	PS-SA-SC
pH	5.3–11 [a]	8.7–12.5 [a]	8.3	7.15	6.59	6.79
Conductivity (mS/cm)	2.5–8.5 [a]	2–30 [a]	2.9	1.6	1.4	1.25
COD (mg/L)	330–1550 [a]	280–7900 [a]	1330	380	405	340
TOC (mg/L)	150–390 [a]	-	180	39	43	32
BOD (mg/L)	1350–1910 [b,c,d]	18–152 [b,c,d]	190	19	21	15
Turbidity (NTU)	36–224 [b,c,d]	1.9–153 [b,c,d]	211	25	29	18

[a] Dilaver et al. (2018) [1]; [b] Qin et al. (2007) [52]; [c] Lopes et al. (2005) [50]; [d] Kadirvelu et al. (2000) [51].

The validity of the model was evaluated by ANOVA [36,37]. The calculated data revealed that significance exceeded 99%, values of the R^2 approached 1, and the Fisher-ratio >> 1. These outcomes attested that the considered model fitted well with the variations of the studied properties versus processing factors.

The equations expressing the change in the density (Y_1), firing shrinkage (Y_2), water absorption (Y_3), and the bending strength (Y_4) according to coded variables, are as follows:

$$Y_1 = 2.084 - 0.018X_1 + 0.001X_2 - 0.03X_3 - 0.01X_1^2 + 0.022X_2^2 + 0.037X_3^2 - 0.07X_1X_2 \\ -0.057X_1X_3 + 0.49X_2X_3 \quad (5)$$

$$Y_2 = 1.531 + 0.89X_1 + 0.462X_2 - 0.065X_3 - 0.006X_1^2 + 0.318X_2^2 - 0.195X_3^2 - 0.225X_1X_2 \\ + 0.109X_1X_3 + 0.449X_2X_3 \tag{6}$$

$$Y_3 = 22.046 + 3.55X_1 - 1.267X_2 + 0.334X_3 - 2.269X_1^2 - 0.056X_2^2 - 0.851X_3^2 - 0.262X_1X_2 \\ -0.022X_1X_3 + 0.204X_2X_3 \tag{7}$$

$$Y_4 = 1.443 - 0.2X_1 + 0.254X_2 - 0.01X_3 - 0.081X_1^2 + 0.286X_2^2 + 0.141X_3^2 + 0.11X_1X_2 \\ -0.045X_1X_3 + 0.409X_2X_3 \tag{8}$$

Examination of linear coefficients values showed that:

- The weights of the effects of the factors studied on the studied properties follows the order: $t > \tau > T$ for density, $\tau > T > t$ for firing shrinkage and water absorption, while for resistance compression, it follows the order $T > \tau > t$.
- Increasing the temperature (T) had a positive effect on all physical properties. In fact, following the increase in temperature, sintering is initiated, and the matrix is consolidated and therefore the mechanical properties are improved.
- The addition of sawdust (τ) had a marked effect on the properties. This effect may be related to the abundance of pores. These were replenished following the decomposition of sawdust, and release CO_2.
- Increasing the soaking time (t) had a positive effect on the density and the shrinkage firing (Figure 14); however, it adversely affects the other two properties. Probably, the adoption of long soaking time favors diffusion phenomena, which leads to an increase in the amount of gehlenite formed, and thus the porosity.
- The effect of interactions between two experimental factors considered changed according to the property. For example, considering the equation of the bending strength, sawdust addition and time have an antagonistic interaction. In other words, the simultaneous increase of these two factors decreases the mechanical strength (Figure 13). The same happens in the case of firing shrinkage for sawdust rate and temperature factors (Figure 14). However, the interaction between the temperature and the soaking time was synergistic in the case of bending strength. This means that the simultaneous increase of T and t led to the formation of mechanically resistant samples (Figure 13).

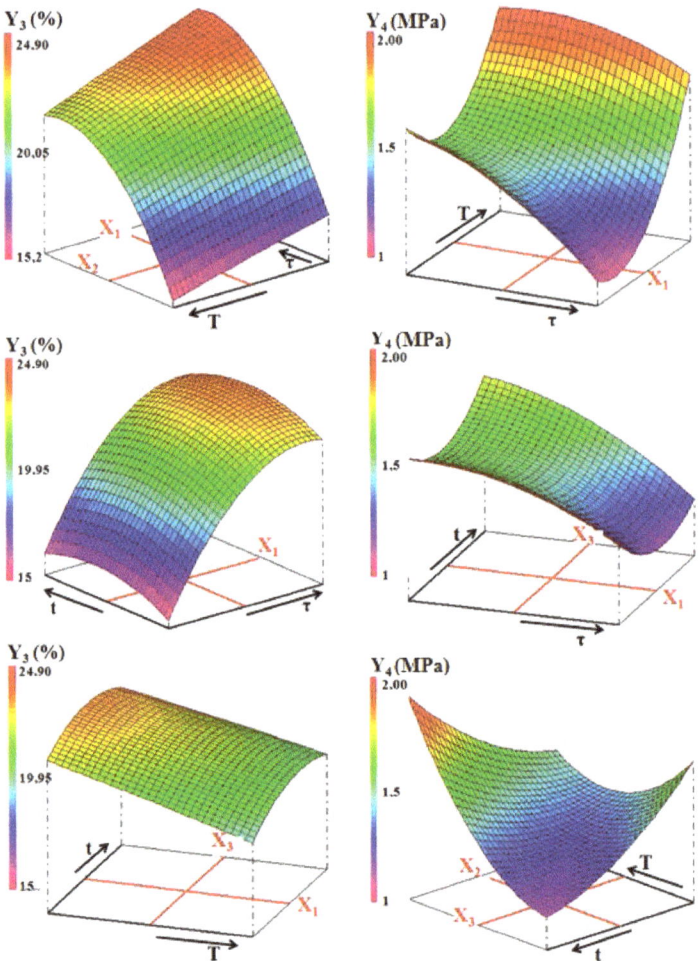

Figure 13. Variation of the water absorption (Y_3) and compressive strength (Y_4) of the ternary blends.

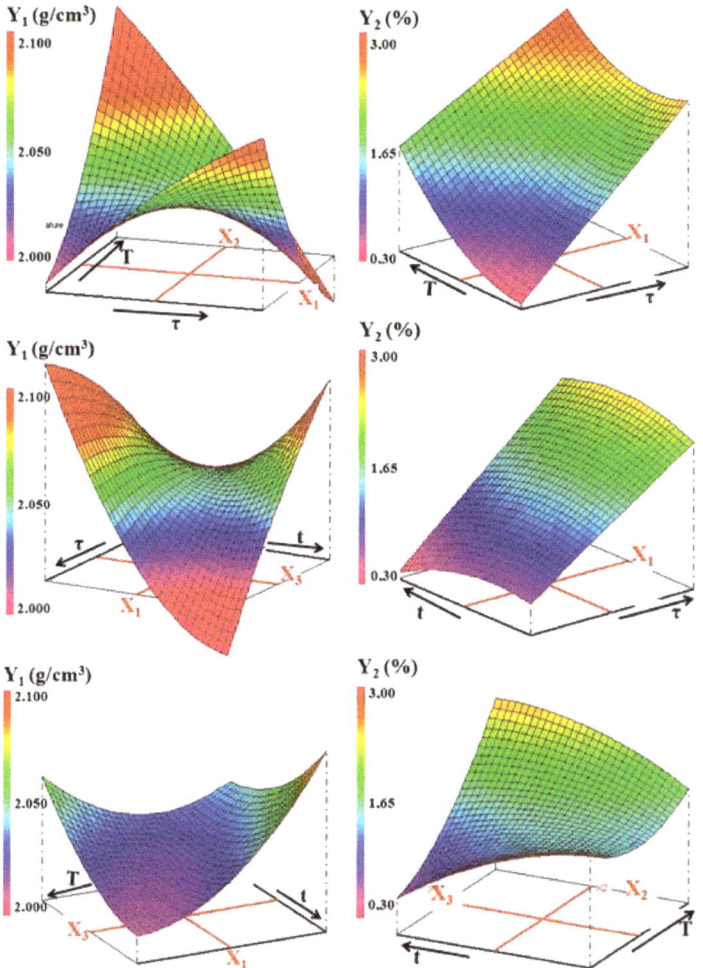

Figure 14. Three-dimensional representation of the variations of density (Y_1) and the firing shrinkage (Y_2) against the processing factors for the ternary mixture.

4. Conclusions

This study reported the manufacturing and the characterization of new ceramic filtration membranes from micaceous clay and phosphate sludge. The results of mineralogical, mechanical, and physical characterizations on the studied materials allowed the conclusions as follows:

- Gehlenite and calcium phosphate neoformed in heated blends containing phosphate sludge (binary and ternary mixture), while the SA-SC mixture was the subject of formation gehlenite and the diopside simultaneously. These neoformed phases were derived from carbonate-released lime and decomposition products of clay minerals.
- It is possible, by adding sawdust, to produce porous bodies.
- The melt formed at high temperature can be connected to the partial melting of fluorapatite.

- Phosphate sludge-based blends showed low mechanical properties (bending strength) compared to clay-based mixtures. That suggests that the incorporation of micaceous clay overcomes this problem because of its high aluminosilicate level.
- The results of filtration tests have shown that the membrane filters based on clay-amended phosphate sludge can be employed in the treatment of textile effluents. In addition, these membrane materials may be used as a carrier of the microfiltration membrane.
- The use of the experimental design allowed assessment of the weight of the effects of experimental factors on the physical properties. Firing temperature and sawdust addition are the most influential factors. Temperature had a positive effect on the studied properties, while sawdust addition has a mitigated effect.

Author Contributions: Conceptualization, M.L.; Data curation, M.L. and W.M.; Formal analysis, M.L., M.K., M.M., L.G. and C.F.; Funding acquisition, R.H.; Investigation, M.L. and M.K.; Methodology, M.L. and R.H.; Project administration, R.H.; Resources, L.G. and Y.T.; Software, M.L. and W.M.; Supervision, M.M., C.F. and R.H.; Validation, M.L.; Writing—original draft, M.L.; Writing—review & editing, R.H.

Funding: This research received no external funding.

Acknowledgments: This work was supported in part by the Erasmus Mundus Program (BATTUTA). Special thanks to the Moroccan Center for Analysis and Characterization (CAC) associated with Cadi Ayyad University (UCA), for helping in some sample characterization.

Conflicts of Interest: The authors declare no conflict of interest.

References

1. Dilaver, M.; Hocaoğlu, S.M.; Soydemir, G.; Dursun, M.; Keskinler, B.; Koyuncu, İ.; Ağtaş, M. Hot wastewater recovery by using ceramic membrane ultrafiltration and its reusability in textile industry. *J. Clean. Prod.* **2018**, *171*, 220–233. [CrossRef]
2. Guo, Y.; Song, Z.; Xu, B.; Li, Y.; Qi, F.; Croue, J.-P.; Yuan, D. A novel catalytic ceramic membrane fabricated with CuMn2O4 particles for emerging UV absorbers degradation from aqueous and membrane fouling elimination. *J. Hazard. Mater.* **2018**, *344*, 1229–1239. [CrossRef] [PubMed]
3. Palacio, L.; Bouzerdi, Y.; Ouammou, M.; Albizane, A.; Bennazha, J.; Hernández, A.; Calvo, J.I. Ceramic membranes from Moroccan natural clay and phosphate for industrial water treatment. *Desalination* **2009**, *245*, 501–507. [CrossRef]
4. Stylianou, S.K.; Katsoyiannis, I.A.; Mitrakas, M.; Zouboulis, A.I. Application of a ceramic membrane contacting process for ozone and peroxone treatment of micropollutant contaminated surface water. *J. Hazard. Mater.* **2018**, *358*, 129–135. [CrossRef] [PubMed]
5. Torres, J.J.; Rodriguez, N.E.; Arana, J.T.; Ochoa, N.A.; Marchese, J.; Pagliero, C. Ultrafiltration polymeric membranes for the purification of biodiesel from ethanol. *J. Clean. Prod.* **2017**, *141*, 641–647. [CrossRef]
6. Dong, Y.; Ma, L.; Tang, C.Y.; Yang, F.; Quan, X.; Jassby, D.; Zaworotko, M.J.; Guiver, M.D. Stable Superhydrophobic Ceramic-Based Carbon Nanotube Composite Desalination Membranes. *Nano Lett.* **2018**, *18*, 5514–5521. [CrossRef]
7. Loutou, M.; Hajjaji, M.; Babram, M.A.; Mansori, M.; Favotto, C.; Hakkou, R. Phosphate sludge-based ceramics: Microstructure and effects of processing factors. *J. Build. Eng.* **2017**, *11*, 48–55. [CrossRef]
8. Loutou, M.; Hajjaji, M.; Mansori, M.; Favotto, C.; Hakkou, R. Phosphate sludge: Thermal transformation and use as lightweight aggregate material. *J. Environ. Manag.* **2013**, *130*, 354–360. [CrossRef]
9. Chrysochoou, M.; Dermatas, D.; Grubb, D.G. Phosphate application to firing range soils for Pb immobilization: The unclear role of phosphate. *J. Hazard. Mater.* **2007**, *144*, 1–14. [CrossRef] [PubMed]
10. Yin, H.; Yun, Y.; Zhang, Y.; Fan, C. Phosphate removal from wastewaters by a naturally occurring, calcium-rich sepiolite. *J. Hazard. Mater.* **2011**, *198*, 362–369. [CrossRef]
11. Lath, S.; Knight, E.R.; Navarro, D.A.; Kookana, R.S.; McLaughlin, M.J. Sorption of PFOA onto different laboratory materials: Filter membranes and centrifuge tubes. *Chemosphere* **2019**, *222*, 671–678. [CrossRef]
12. Roussanaly, S.; Anantharaman, R.; Lindqvist, K.; Hagen, B. A new approach to the identification of high-potential materials for cost-efficient membrane-based post-combustion CO_2 capture. *Sustain. Energy Fuels* **2018**, *2*, 1225–1243. [CrossRef]

13. Abbasi, M.; Mirfendereski, M.; Nikbakht, M.; Golshenas, M.; Mohammadi, T. Performance study of mullite and mullite–alumina ceramic MF membranes for oily wastewaters treatment. *Desalination* **2010**, *259*, 169–178. [CrossRef]
14. Wan, W.; Zhang, R.; Ma, M.; Zhou, Y. Monolithic aerogel photocatalysts: a review. *J. Mater. Chem. A* **2018**, *6*, 754–775. [CrossRef]
15. Fontes, W.C.; Franco de Carvalho, J.M.; Andrade, L.C.R.; Segadães, A.M.; Peixoto, R.A.F. Assessment of the use potential of iron ore tailings in the manufacture of ceramic tiles: From tailings-dams to "brown porcelain". *Constr. Build. Mater.* **2019**, *206*, 111–121. [CrossRef]
16. Loutou, M.; Hajjaji, M.; Mansori, M.; Favotto, C.; Hakkou, R. Heated blends of clay and phosphate sludge: Microstructure and physical properties. *J. Asian Ceram. Soc.* **2016**, *4*, 11–18. [CrossRef]
17. Loutou, M.; Hajjaji, M.; Mansori, M.; Favotto, C.; Hakkou, R. Heated blends of phosphate waste: Microstructure characterization, effects of processing factors and use as a phosphorus source for alfalfa growth. *J. Environ. Manag.* **2016**, *177*, 169–176. [CrossRef]
18. Loutou, M.; Hajjaji, M. Clayey wastes-based lightweight aggregates: Heating transformations and physical/mechanical properties. *Appl. Clay Sci.* **2017**, *150*, 56–62. [CrossRef]
19. Kang, D.-Y.; Lydon, M.E.; Yucelen, G.I.; Jones, C.W.; Nair, S. Inside Cover: Solution-Processed Ultrathin Aluminosilicate Nanotube–Poly(vinyl alcohol) Composite Membranes with Partial Alignment of Nanotubes (ChemNanoMat 2/2015). *ChemNanoMat* **2015**, *1*, 70. [CrossRef]
20. Lü, Q.; Dong, X.; Zhu, Z.; Dong, Y. Environment-oriented low-cost porous mullite ceramic membrane supports fabricated from coal gangue and bauxite. *J. Hazard. Mater.* **2014**, *273*, 136–145. [CrossRef]
21. Rasouli, Y.; Abbasi, M.; Hashemifard, S.A. Investigation of in-line coagulation-MF hybrid process for oily wastewater treatment by using novel ceramic membranes. *J. Clean. Prod.* **2017**, *161*, 545–559. [CrossRef]
22. Silva, R.V.; de Brito, J.; Lye, C.Q.; Dhir, R.K. The role of glass waste in the production of ceramic-based products and other applications: A review. *J. Clean. Prod.* **2017**, *167*, 346–364. [CrossRef]
23. Almandoz, M.C.; Pagliero, C.L.; Ochoa, N.A.; Marchese, J. Composite ceramic membranes from natural aluminosilicates for microfiltration applications. *Ceram. Int.* **2015**, *41*, 5621–5633. [CrossRef]
24. Qiu, H.; Jiang, J.; Peng, L.; Liu, H.; Gu, X. Choline chloride templated CHA zeolite membranes for solvents dehydration with improved acid stability. *Microporous Mesoporous Mater.* **2019**, *284*, 170–176. [CrossRef]
25. Funk, J.E.; Dinger, D.R. *Predictive Process Control of Crowded Particulate Suspensions: Applied to Ceramic Manufacturing*; Springer Science & Business Media: Berlin, Germany, 2013; ISBN 978-1-4615-3118-0.
26. Fernandes, H.R.; Tulyaganov, D.U.; Ferreira, J.M.F. Preparation and characterization of foams from sheet glass and fly ash using carbonates as foaming agents. *Ceram. Int.* **2009**, *35*, 229–235. [CrossRef]
27. Živcová, Z.; Gregorová, E.; Pabst, W.; Smith, D.S.; Michot, A.; Poulier, C. Thermal conductivity of porous alumina ceramics prepared using starch as a pore-forming agent. *J. Eur. Ceram. Soc.* **2009**, *29*, 347–353. [CrossRef]
28. Živcová, Z.; Černý, M.; Pabst, W.; Gregorová, E. Elastic properties of porous oxide ceramics prepared using starch as a pore-forming agent. *J. Eur. Ceram. Soc.* **2009**, *29*, 2765–2771. [CrossRef]
29. Cao, J.; Rambo, C.R.; Sieber, H. Preparation of Porous Al2O3-Ceramics by Biotemplating of Wood. *J. Porous Mater.* **2004**, *11*, 163–172. [CrossRef]
30. Sutcu, M.; Akkurt, S. The use of recycled paper processing residues in making porous brick with reduced thermal conductivity. *Ceram. Int.* **2009**, *35*, 2625–2631. [CrossRef]
31. Bose, S.; Das, C. Sawdust: From wood waste to pore-former in the fabrication of ceramic membrane. *Ceram. Int.* **2015**, *41*, 4070–4079. [CrossRef]
32. Sutcu, M.; Akkurt, S.; Bayram, A.; Uluca, U. Production of anorthite refractory insulating firebrick from mixtures of clay and recycled paper waste with sawdust addition. *Ceram. Int.* **2012**, *38*, 1033–1041. [CrossRef]
33. Eliche-Quesada, D.; Corpas-Iglesias, F.A.; Pérez-Villarejo, L.; Iglesias-Godino, F.J. Recycling of sawdust, spent earth from oil filtration, compost and marble residues for brick manufacturing. *Constr. Build. Mater.* **2012**, *34*, 275–284. [CrossRef]
34. Cook, R.A.; Hover, K.C. Mercury porosimetry of hardened cement pastes. *Cem. Concr. Res.* **1999**, *29*, 933–943. [CrossRef]

35. Saffaj, N.; Persin, M.; Younsi, S.A.; Albizane, A.; Cretin, M.; Larbot, A. Elaboration and characterization of microfiltration and ultrafiltration membranes deposited on raw support prepared from natural Moroccan clay: Application to filtration of solution containing dyes and salts. *Appl. Clay Sci.* **2006**, *31*, 110–119. [CrossRef]
36. Mora-Tamez, L.; Barim, G.; Downes, C.; Williamson, E.M.; Habas, S.E.; Brutchey, R.L. Controlled Design of Phase- and Size-Tunable Monodisperse Ni2P Nanoparticles in a Phosphonium-Based Ionic Liquid through Response Surface Methodology. *Chem. Mater.* **2019**, *31*, 1552–1560. [CrossRef]
37. Njoya, D.; Hajjaji, M. Quantification of the effects of manufacturing factors on ceramic properties using full factorial design. *J. Asian Ceram. Soc.* **2015**, *3*, 32–37. [CrossRef]
38. Pińkowska, H.; Krzywonos, M.; Wolak, P.; Złocińska, A. Pectin and Neutral Monosaccharides Production during the Simultaneous Hydrothermal Extraction of Waste Biomass from Refining of Sugar—Optimization with the Use of Doehlert Design. *Molecules* **2019**, *24*, 472. [CrossRef]
39. Ben Khalifa, E.; Rzig, B.; Chakroun, R.; Nouagui, H.; Hamrouni, B. Application of response surface methodology for chromium removal by adsorption on low-cost biosorbent. *Chemom. Intell. Lab. Syst.* **2019**, *189*, 18–26. [CrossRef]
40. Yurekli, Y.; Yildirim, M.; Aydin, L.; Savran, M. Filtration and removal performances of membrane adsorbers. *J. Hazard. Mater.* **2017**, *332*, 33–41. [CrossRef] [PubMed]
41. Šereš, Z.; Maravić, N.; Takači, A.; Nikolić, I.; Šoronja-Simović, D.; Jokić, A.; Hodur, C. Treatment of vegetable oil refinery wastewater using alumina ceramic membrane: optimization using response surface methodology. *J. Clean. Prod.* **2016**, *112*, 3132–3137. [CrossRef]
42. Vatanpour, V.; Karami, A.; Sheydaei, M. Improved visible photocatalytic activity of TiO2 nanoparticles to use in submerged membrane photoreactor for organic pollutant degradation. *Int. J. Environ. Sci. Technol.* **2019**, *16*, 2405–2414. [CrossRef]
43. Valverde, J.M.; Perejon, A.; Medina, S.; Perez-Maqueda, L.A. Thermal decomposition of dolomite under CO2: insights from TGA and in situ XRD analysis. *Phys. Chem. Chem. Phys.* **2015**, *17*, 30162–30176. [CrossRef]
44. Gunasekaran, S.; Anbalagan, G. Thermal decomposition of natural dolomite. *Bull. Mater. Sci.* **2007**, *30*, 339–344. [CrossRef]
45. Eliche-Quesada, D.; Martínez-Martínez, S.; Pérez-Villarejo, L.; Iglesias-Godino, F.J.; Martínez-García, C.; Corpas-Iglesias, F.A. Valorization of biodiesel production residues in making porous clay brick. *Fuel Process. Technol.* **2012**, *103*, 166–173. [CrossRef]
46. Xie, J.; Chen, T.; Xing, B.; Liu, H.; Xie, Q.; Li, H.; Wu, Y. The thermochemical activity of dolomite occurred in dolomite–palygorskite. *Appl. Clay Sci.* **2016**, *119*, 42–48. [CrossRef]
47. Qian, H.; Kai, W.; Hongde, X. A novel perspective of dolomite decomposition: Elementary reactions analysis by thermogravimetric mass spectrometry. *Thermochim. Acta* **2019**, *676*, 47–51. [CrossRef]
48. Rahaman, M.N. *Ceramic Processing*; CRC Press: Boca Raton, FL, USA, 2017; ISBN 978-1-315-15716-0.
49. Studart, A.R.; Gonzenbach, U.T.; Tervoort, E.; Gauckler, L.J. Processing Routes to Macroporous Ceramics: A Review. *J. Am. Ceram. Soc.* **2006**, *89*, 1771–1789. [CrossRef]
50. Lopes, C.N.; Petrus, J.C.C.; Riella, H.G. Color and COD retention by nanofiltration membranes. *Desalination* **2005**, *172*, 77–83. [CrossRef]
51. Kadirvelu, K.; Palanival, M.; Kalpana, R.; Rajeswari, S. Activated carbon from an agricultural by-product, for the treatment of dyeing industry wastewater. *Bioresour. Technol.* **2000**, *74*, 263–265. [CrossRef]
52. Qin, J.-J.; Oo, M.H.; Kekre, K.A. Nanofiltration for recovering wastewater from a specific dyeing facility. *Sep. Purif. Technol.* **2007**, *56*, 199–203. [CrossRef]
53. Van der Bruggen, B.; Daems, B.; Wilms, D.; Vandecasteele, C. Mechanisms of retention and flux decline for the nanofiltration of dye baths from the textile industry. *Sep. Purif. Technol.* **2001**, *22–23*, 519–528. [CrossRef]

© 2019 by the authors. Licensee MDPI, Basel, Switzerland. This article is an open access article distributed under the terms and conditions of the Creative Commons Attribution (CC BY) license (http://creativecommons.org/licenses/by/4.0/).

Article

Valorization of Phosphate Mine Waste Rocks as Materials for Road Construction

Mustapha Amrani [1,2,*], Yassine Taha [3,*], Azzouz Kchikach [1], Mostafa Benzaazoua [4,5] and Rachid Hakkou [2,3]

1. L3G Laboratory, équipe de recherche Génie Civil et Géo-Ingénierie, Univ. Cadi Ayyad (UCA), BP 549, Marrakech 40000, Morocco; kchikach@gmail.com
2. LCME, Faculté des Sciences et Techniques, Univ. Cadi Ayyad, BP 549, Marrakech 40000, Morocco; r.hakkou@uca.ma
3. Materials Science and Nano-Engineering Department, Mohammed VI Polytechnic University, Lot 660, Hay Moulay Rachid, Ben Guerir 43150, Morocco
4. Institut de Recherche en Mines et en Environnement, Univ. du Québec en Abitibi Témiscamingue, 445 Boul de l'Université, Rouyn-Noranda, QC J9X 5E4, Canada; mostafa.benzaazoua@uqat.ca
5. Geology and Sustainable Mining Department, Mohammed VI Polytechnic University, Lot 660, Hay Moulay Rachid, Ben Guerir 43150, Morocco
* Correspondence: amrani.be.2p@gmail.com (M.A.); yassine.taha@um6p.ma (Y.T.); Tel.: +212-661-333-542 (M.A.); +212-525-072-867 (Y.T.)

Received: 24 February 2019; Accepted: 15 April 2019; Published: 17 April 2019

Abstract: The road construction sector is a worldwide high consumer of natural aggregates. The use of unusual industrial by-products in road techniques can contribute to the conservation of non-renewable natural resources and the reduction of wastes produced by some industries. Phosphate waste rocks could be considered as potential alternative secondary raw materials in road construction. The use and valorization of these wastes is currently limited according to the Moroccan guide for road earthworks (GMTR). The guide has classified these materials as waste products, which consequently, cannot be used in road construction. However, phosphate waste rocks are sedimentary natural rocks which have not been subjected to any transformation other than mechanical fragmentation. The goal of this paper is to discuss key-properties of various phosphate mine waste rocks (PMWR) to be used as road materials. Samples were taken from different stockpiles in the phosphate mine site of Gantour in Morocco. The different waste rocks samples were characterized in terms of their physical, geotechnical, chemical, mineralogical and environmental properties using international testing norms. The obtained results showed that the studied PMWR presented satisfying characteristics; the specific (particle) density: $\rho s > 26$ kN/m^3, Los Angeles abrasion: 45% < LA < 58%), methylene blue value MBV < 1 g/100g, organic matter: OM < 1% and plasticity index: PI < 20%. All PMWR were confirmed as possessing the requested geotechnical properties to be used as materials for embankments. Moreover, leaching tests showed that none of them released any contaminants. In field application, these materials have been also successfully used in in situ experimental pilot testing. Therefore, the PMWR have to be classified in the category of natural aggregates that are similar to conventional materials.

Keywords: civil engineering; valorization; phosphate mine waste rocks; natural aggregates; road techniques wet process

---

1. Introduction

Various stakeholders in the road construction sector have to deal with the increased demand of raw materials used in road infrastructures. The flexible pavement structure is generally composed of several layers of materials: embankment materials, capping layer, pavement aggregate (base and sub-base course) and surfacing (surface and binder course) course.

Recently, an increasing attention has been given to the potential use of alternative aggregates, particularly, in the road construction sector [1]. Several factors must be considered before using a waste or by-product in road engineering. The use of industrial wastes as secondary and alternative materials in the infrastructure sector depends on their availability, on the transport costs and their physical, geotechnical and chemical properties. The toxicity and solubility in water is a relevant factor to be considered with other factors [2]. The use of alternative materials in road construction provides several economic and ecological advantages. When waste materials with acceptable properties are available, it is possible to avoid the costs related to the extraction, and to minimize the transport distance, energy consumption and consequently the greenhouse gas emissions [3].

Many examples have been studied in detail in the literature. Fly ash and other agricultural wastes were used as soil admixture to improve the CBR values of soil in lower layers of road construction [4]. Incinerated bottom ashes were also investigated in road construction [5–7]. When stabilized with binder additives, these materials could be used successfully in embankment and pavement layers. Construction or demolition wastes have been used in the construction of embankments and pavement layers [2,8,9]. It was also demonstrated that steel-slag fly ash and phosphogypsum as a solidified material can be used as road materials with competitive characteristics [10–12]. It has been established that the use of fly ashes could improve the natural and mechanical characteristics of soils [13]. Dredged sediments mixed with binders (cement and/or lime) and other products (steel slag, fly ash) were compatible with the requested standards for their use as base or embankments course material [14–17]. According to the inventory carried out by OCDE, about twenty types of waste and by-products, to be used in road engineering, has been studied [1]. A classification according to the origin, the main characteristics, the current and the potential uses has been proposed.

In Morocco, phosphate mines produce millions of tons of phosphate mine waste rocks (PMWR) which are stockpiled on surface in waste rock piles covering large areas (several thousand hectares). These waste rocks represent mainly the intercalation layers (limestones, marls and flintstone) and the cover layer (topsoil, clays and marls) occurring within the phosphate sequence. During the extraction of phosphate ore, intercalation layers and the cover layer are blasted and stripped away. Due to their high calcite and dolomite content, the PMWR are inert geochemically. Hakkou, et al. [18] demonstrated that PMWR could be used as materials for the passive treatment of acid mine drainage [19,20], and had the appropriate properties for a store-and-release cover component in a semiarid climate [21]. Although they have characteristics similar to the natural aggregates used as building materials, PMWR are classified by the Moroccan Guide for Road Earthworks [22] in the organic soils and industrial by-products class and particularly in the phosphate wastes sub-class F3 and cannot thus, be used in road construction.

To our knowledge, very limited scientific research on the valorization of PMWR in road construction has been published. Ahmed and Abouzeid [23] investigated the use of phosphate waste rocks in road construction. The work consisted of geotechnical characterization, which led them to conclude an interesting potential of these by-products as road aggregates; similar to natural ones. The substitution of conventional aggregates by PMWR for the construction of road infrastructures might be considered as a promising and ecofriendly solution. A scientific approach of valorization of these mining wastes in road technique is, therefore, necessary to ensure the transition from a "waste" to "building materials". The aim of this paper is to focus on PMWR and its use in the construction of road embankments and to discuss their status in Moroccan Guide for Road Earthworks. Laboratory tests were performed in order to determine the chemical, mineralogical characteristics, physical and geotechnical properties and the environmental behavior of the PMWR. In addition, in situ tests were realized in order to access the behavior of these materials during and after embankments construction using the wet process.

2. Materials and Methods

2.1. Materials Sampling

The studied materials were sampled from the mining site located in the central part of the sedimentary phosphate deposits in the Guantour region (Figure 1). The deposit is characterized by phosphate series of late Cretaceous–Eocene age consisting of alternating layers of phosphate separated by gangue silico-carbonate levels. The upper phosphate layer is overburdened by an alternation of different layers (topsoil, siliceous marl, clays, flintstone, calcareous marl and alluvium). The mine produces millions of tons per year of phosphate mine waste rocks (PMWR). The various rock lithologies are scattered in the mine site. Five different waste rocks piles referenced hereafter as I1, I2, I3, I4 and I5 were investigated. To ensure a representative sample, attention was given to the mode, history of storage, geological and petrographic lithology description to identify the parameters likely to impact the characteristics of the sampled materials. Given the heterogeneity of the waste rock piles, a rigorous technique was used to obtain the most representative sampling. The approach consisted of collecting 3 samples from the five stockpiled waste rocks of approximately equal size at different points, respectively at the base, at the middle and at the top of the pile depending on the actual segregation status (Figure 2). The field samples were collected, homogenized in the laboratory, and riffle-split into smaller sub-samples which were stored for further testing.

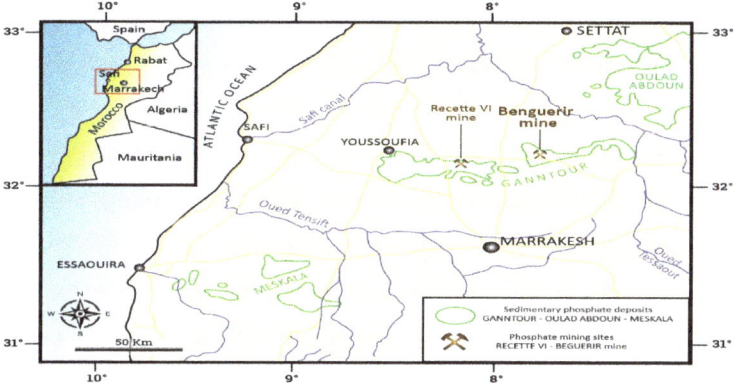

Figure 1. Geographical location of the studied mine site.

Figure 2. Phosphate mine waste rocks dumps.

2.2. Research Methodology

The GMTR guide does not include PMWR specifications and conditions of use, therefore, the geotechnical characterization in the laboratory has been completed by conducting in situ tests in a trail embankment and identifying other specific parameters (consolidation, shear strength, chemical and environmental properties) which are not provided by the NF P11-300 standard and which may affect their functional behavior. Figure 3 highlights a summary of the methodology used in this work.

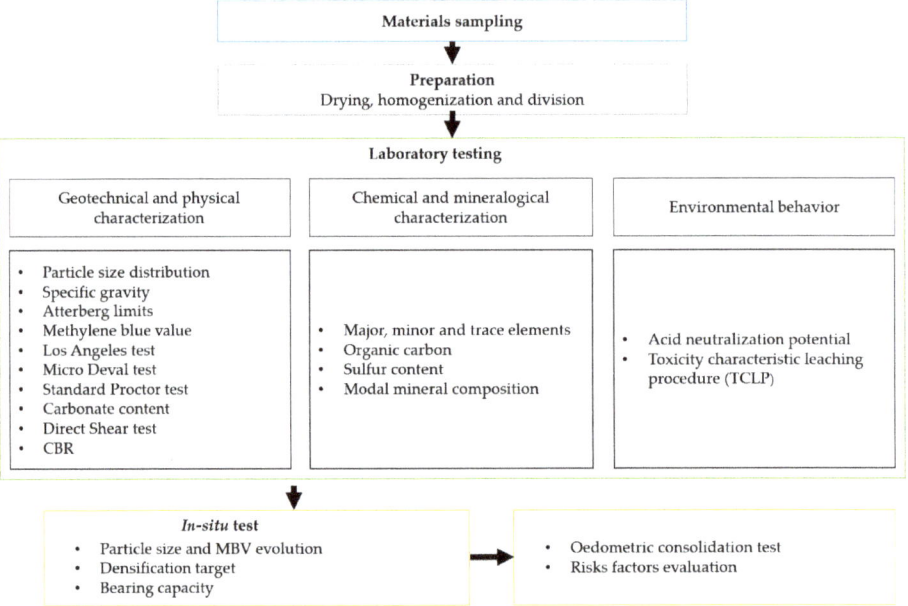

Figure 3. Summary of the followed methodology.

2.2.1. Laboratory Tests

Laboratory tests have been conducted to determine the chemical characteristics, the environmental behavior, the mechanical and physical properties and the geotechnical properties. The laboratory tests were carried out in accordance with the relevant AFNOR standards [24]. The leaching behavior of waste rocks was assessed using the Toxicity Characteristic Leaching Procedure (TCLP) [25]. The leaching solution used has a pH of 4.93 ± 0.05. The obtained results are compared with US-EPA thresholds [26]. This test is used to verify a potential release of impurities and contaminants. The chemical composition of the solid samples was analyzed using an X-ray Fluorescence (Panalytical, Epsilon 4 Model, Malvern Panalytical, Malvern, UK) and liquid solutions were analyzed by inductively coupled plasma with atomic emission spectroscopy (ICP-AES) (Perkin Elmer Optima 3100 RL, PerkinElmer Waltham, MA, USA). The organic carbon content (Corg) was determined by dichromate oxidation in the presence of concentrated sulphuric acid according to ISO 14235 standard [27]. The crystalline phases were determined by the X-ray diffraction (Bruker AXS Advance D8. Bruker, Billerica, MA, USA), Cu Kα radiation. The DiffracPlus EVA software (Bruker, Billerica, MA, USA) was used to identify mineral species and TOPAS software (Bruker, Billerica, MA, USA) implementing Rietveldt refinement to quantify the abundance of all identified minerals. Due to the presence of calcareous rocks, the carbonate content was determined on −400 μm of crushed samples in accordance with the NF P 94-048 standard [28].

The classification of the PMWR has been codified in relation to the nearest standardized materials, it will be therefore called to standard NF P 11-300 [24] for the geotechnical characterization. Prior to the characterization, the moisture content of the samples was measured [29]. The fraction 0/400 μm of the studied samples have been subjected to the plasticity test using the Atterberg limits method [30,31]. Also, dimensional properties tests were investigated using granularity method by dry sieving after washing [32]. The methylene blue test value (MBV) was measured on the 0/5-mm fraction taken from the 0/50-mm dry material according to the standard NF P 94-068 [33]. For compaction aptitude, the samples were compacted, in three layers, in a CBR standard mold using normal compaction [34]. When the proportion of elements greater than 20 mm exceeds 30% of the mass of the material, the Proctor test was performed on the 0/20-mm fraction, but its interpretation is then limited to the assessment of its moisture content w_{opn}. In this case, the real dry density was measured in a full trial scale. To complete the knowledge of the petrographic features of the original rocks, evaluate the resistance of the material regarding the fragmentation, the wear and the particle size distribution evolution under the effect of mechanical solicitations, several tests were carried out using: NF P 94-064 standard for the density of a rock element by the hydrostatic weighing method. For better representativeness, the fraction 25/50 mm was chosen for the determination of Los Angeles abrasion value (LA) and Micro Deval value (MD) using NF P18-573 [35] and P18-572 [36] standards. To measure the sensitivity of these materials to fragmentation under the effect of mechanical stresses and climatic cycles, the fragmentation coefficient and degradability coefficient of samples were measured on the 40/80-mm fraction according to the standards NF-P94-066 [37] and NF-P94-067 [38], respectively. To determine the load bearing capacity of the materials after compaction and to assess their resistance to punching and heavy machines traffic, Californian bearing ratio (CBR) tests were carried out according to standard NF-P94-078 [39]. The shear strength parameters were investigated under the drained conditions, on the 0/20-mm fraction of the studied waste rock samples on the waste rocks samples [40].

In order to evaluate the vertical deformation (settlement) under the effect of the charges after saturation, an oedometric test was carried out on the 0/20-mm fraction of material I2 according to the standard XP-P94-090-1 [41] with a water content close to the optimum moisture content (w_{opn}) and with a dry density equal to the reference dry density (γdr) determined by the in situ tests.

2.2.2. In Situ Tests

In situ tests were performed with sample I2 which is the most abundant material among the PMWR. The objective was to determine the optimal conditions that offer the best results in terms of material compaction. These are related to: water content, thickness of compacted layer, speed and number of compactor passes. The ability of these materials to achieve the targeted compaction level for the embankments (q4 level) was also assessed. The dry densities of the materials were measured in situ using a membrane densitometer [42], while the measurement of the bearing capacity of the materials was evaluated through the determination of the module under static loading at the plate EV_2 (standard NF P 94-177-1).

A full-scale of trial embankment was constructed on 30 m length and 8 m width reinforced with around 1 m of embankment high. Three layers of 0.30-m thickness were constructed by wet method using a loading machine, a grader, a sprinkler vehicle and vibratory road roller (Figure 4). The trial test was carried out on a stable platform whose bearing capacity at the time of completion was greater than 50 MPa. The layers were compacted with a calibrated vibratory road roller with a single drum [43]. The speed was fixed at 4 km/h and the compaction energy of the machine was controlled. The representation of the three layers pile is shown in Figure 4. Many parameters were fixed during these tests: the speed of the compactor, the vibration amplitude of the compactor, the average water content range (between 0.9 and 1.1 of w_{opn}) and the layer thickness.

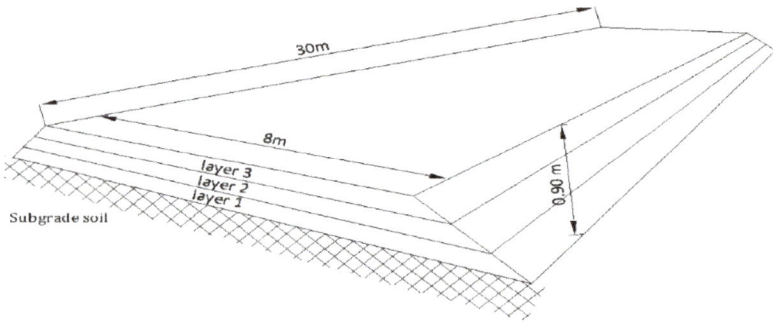

Figure 4. Photos and scheme illustrating the field trial embankment construction.

The shape of the granulometric curve obtained before compacting showed a continuity of the granularity, in addition, to have a better representativity, the water content was realized on the entire fraction of soil-rockfill mixture (0/80 mm).

Before compacting, the grain-size distribution curve, the MBV, the fine fraction (having a size less than 80 µm), the particles having a size less than 2 mm (determined on the 0/50-mm fraction) and the water content were measured. In addition to the same parameters followed before compacting, the layer thickness, the dry density and the bearing capacity for the three compacted layers with the different energy proposed (2, 4 and 8 of roller passes) were measured after compacting. The average of six samples will be selected for each monitored parameter before and after compaction.

The optimum dry density will be the one that corresponds to the maximum value of the six points of dry densities recorded on the compacted thickness and the control of the homogeneity of the distribution of compaction forces was verified by determining the maximum deviation from the mean value on the same compacted layer. In other words, the determination of the compaction energy which makes it possible to have the maximum dry density. It is this which will be taken as reference dry density (γdr) for the calculation of the compaction rate and evaluation the ability of these materials to achieve the required compaction levels for embankments.

3. Results and Discussion

3.1. PMWR Characterization

3.1.1. Physical and Geotechnical Properties

Results of the geotechnical identification tests are summarized in Table 1. All the studied PMWR samples display approximately the same water content and very dry moisture content due to the arid climate of the region and the storage at the mine site. In addition, the tested materials show generally similar properties with a slight difference in grain size distribution results (Figure 5). The 0/20-mm fraction is important as it is considered in the Proctor and CBR tests. Only materials whose +20 mm particles weight proportion under 30% have been the subject of CBR and Proctor tests (it is the case only of I3 material). Also, the granulometry has a direct impact on plasticity of the studied materials. The materials with the highest content of fine fraction (I3, I4 and I5) showed the most plasticity features. The degree of plasticity that remains low for all these materials is particularly related to the mineralogical composition of the clays (illite). A difference was founded in mechanical properties of the studied PMWR samples. Unlike other materials with low values, I2 and I1 samples showed satisfactory values of LA, MD, degradability and fragmentability coefficients. The analysis of the results of the various physical tests indicates that the degradation of particles is related particularly to the presence rate of clay and flintstone in the studied samples. With a maximum particles diameter less than 150 mm, a non-zero cohesion (4–7 kPa), a plasticity index less than 16%, a specific (particle) density around 26 kN/m^3 and a percentage of fine elements less than 23%, the PMWR studied as shown in Table 1 could be classified in the category C1B5 (friable soil); a gravelly coherent materials with a fine fraction [24] and the category of mixture of limestone and siliceous sandstone (for the case of the rocky origin). Thus, these materials can be used in the construction of road embankments. This is illustrated in the synoptic table of classification according to the nature of the materials proposed by the NF P11-300 standard (Figure 6).

Table 1. Physical and geotechnical properties of collected materials.

Test		I1	I2	I3	I4	I5
Moisture Content	wt. %	3.4	3.7	3.6	2.9	3.1
Geotechnical properties—Natural parameters						
Proctor Test						
Optimum Moisture content (w_{opn})	wt. %	13.40	12.90	15.20	14.60	13.23
Maximum dry density $\gamma_{d\,max}$	kN/m^3	*	*	17.9	*	*
Shear test						
Friction angle (Ø')	degrees	30.00	32.40	27.00	27.5	27.00
Cohesion (c')	kPa	4	5	6	7	7
CBR	%	*	*	13	*	*
Atterberg limit						
Liquid limit	wt. %	39	37	41	44	45
Plastic limit	wt. %	26	25	26	29	30
Plasticity index	wt. %	13	12	14	15	15
Methylene blue value	g/100g	0.59	0.58	0.67	0.68	0.71
Carbonate content	wt. %	30	29	33	32	33
Geotechnical Properties–Mechanical behavior						
Specific (particle) densi	kN/m^3	2.61	2.65	2.56	2.6	2.58
Los Angeles abrasion test 25/50	wt. %	48	46	66	67	53
Mico Deval test 25/50	wt.%	55	50	68	70	54
Degradability coefficient	wt.%	10.10	9.10	13.80	14.60	12.70
Fragmentability coefficient	wt.%	8.90	7.50	10.10	11.40	10.50
Material classification	-	C_1B_5	C_1B_5	C_1B_5	C_1B_5	C_1B_5

* Proportion of particles greater than 20 mm exceeds 30% of the mass of the material, in this case, the maximum dry density of the proctor and the value of CBR test are not significant.

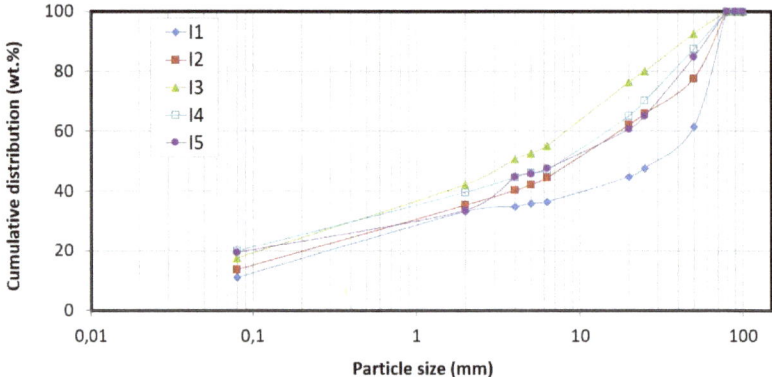

Figure 5. Particle size distribution of the five collected materials.

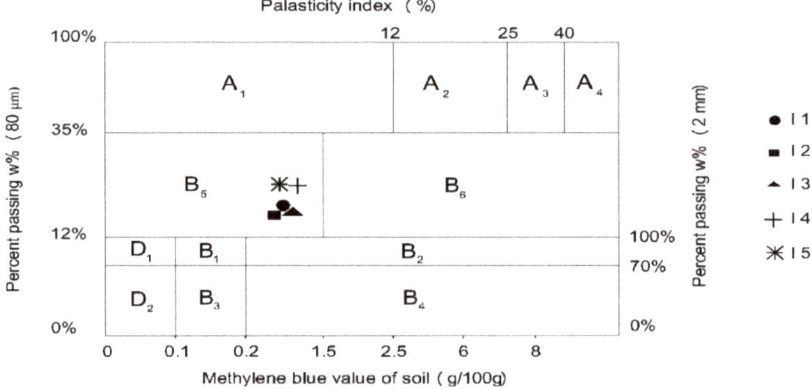

Figure 6. Classification of studied PMWR according to the soil classification table of the NF P11-300 standard (fraction 0/50 mm). A: fine soil; B: Sandy and gravely soil with fine particle and D: Soil insensitive to water.

3.1.2. Chemical and Mineralogical Properties

The chemical and mineralogical composition of PMWR are presented in Table 2. All the samples contain mainly SiO_2 (41–57 wt. %), CaO (12–19 wt. %), MgO (4.9–9.1 wt. %), P_2O_5 (4.2–5.4 wt. %). Alkali and alkali earth oxides are present in low concentrations (less than 1 wt. %). The amounts of detected sulfur and organic carbon were generally below 0.5 wt. %. The PMWR could be classified easily as non-generating of acid mine drainage with a very neutralization potential, as already demonstrated [18]. In terms of trace element occurrence, only a very low concentration of Cr, Cd, Cu and Mo were detected. The relative abundance of minerals identified by XRD and quantified using a Rietveld refinement method is illustrated in Table 2. All materials contain a siliceous fraction represented by quartz and cristobalite, and carbonaceous fraction represented mainly by dolomite and little amount of calcite. Fluorapatite was also detected varying between 6 and 8.5wt. %. The mineralogical composition of the studied materials showed also plagioclase (albite and anorthite) and clays (illite which was observed only in samples I3, I4 and I5, that is why they showed the most plasticity features).

Table 2. Chemical and mineralogical composition of PMWR samples.

PMWR Sample		I1	I2	I3	I4	I5
Major elements (wt. %)						
SiO_2		41.30	50.10	55.50	53.90	56.40
Al_2O_3		0.40	0.44	4.10	3.80	3.10
Fe_2O_3		-	-	0.50	0.40	0.30
CaO		18.70	16.20	12.10	12.90	12.50
MgO		9.10	7.20	4.90	5.50	5.40
K_2O		-	-	1.60	1.31	1.00
P_2O_5		4.22	5.40	5.10	4.70	4.20
LOI		24.20	18.30	15.50	17.40	16.90
C_{org}		0.21	0.18	0.31	0.44	0.38
S		0.30	0.40	0.20	0.30	0.30
Mineralogical composition (wt. %)						
Quartz	SiO_2	41.21	49.80	32.00	30.10	33.60
Cristobalite	SiO_2			17.40	18.40	18.40
Dolomite	$(Ca,Mg)(CO_3)_2$	40.89	32.00	21.00	24.00	24.00
Calcite	$CaCO_3$	6.04	5.00	3.90	3.50	3.70
Fluorapatite	$Ca_5(PO_4)_3F$	6.42	6.20	8.30	8.10	7.90
Albite	$NaAlSi_3O_8$	5.06	6.40	6.20	5.60	4.90
Illite	$(K,H_3O)(Al,Mg)_2(Si,Al)_4O_{10}[(OH)_2,(H_2O)]$			11.00	9.20	6.50
Anorthite	$CaAl_2Si_3O_8$	0.38	0.60	0.44	0.80	0.90

LOI: Loss on Ignition, C_{org}: Organic Carbon.

No montmorillonite has been detected, which eliminates the risk of swelling of the clayey fraction contained in these materials once used in road construction. The I2 sample exhibit low content of clays (illite) and high content of siliceous minerals (flintstone occurrence), which explain the best mechanical features: Los Angeles (LA = 46 wt. %), Micro Deval (MD = 50 wt. %), Degrability and Fragmentability coefficients respectively 9.1 and 7.5 as shown in Table 1.

3.1.3. Environmental Behavior of Materials

The results of trace elements leaching from PMWR using the TCLP test are summarized in Table 3. The mobility of heavy metals depends on several factors such as heavy metals bearing minerals and the pH of the leaching solution. All concentrations were in agreement with the limits for non-hazardous waste fixed by US-EPA regulation. Therefore, the studied PMWR cannot be listed as hazardous waste, in fact they should be considered as natural aggregates. The observed limited metal release is explained by the low initial content within PMWR and the relative high stability of the occurring inert minerals (silice and aluminosilicates) and high neutralizing capacity minerals such as dolomite and calcite. The fluoroapatite need strong acidity to be solubilized.

Table 3. Results of the Toxicity Characteristic Leaching Procedure (TCLP) of PMWR.

Sample	Zn	Se	Pb	Cu	Cr	Cd	As	V
	mg/L	mg/L	mg/L	mg/L	mg/L	mg/L	mg/L	mg/L
I1	0.55	<0.10	<0.60	<0.50	<0.20	<0.10	<1	<1
I2	0.62	<0.10	<0.60	<0.50	<0.20	<0.10	<1	<1
I3	0.73	<0.10	<0.60	<0.50	<0.20	<0.10	<1	<1
I4	0.54	<0.10	<0.60	<0.50	<0.20	<0.10	<1	<1
I5	0.56	<0.10	<0.60	<0.50	<0.20	<0.10	<1	<1
Limits (US-EPA)	2	1	5	-	5	1	5	-

3.2. In situ Full Trial Tests

The evolution criterion (in the granulometric approach) considered in this on-site study concerns fragmentation under the action of mechanical solicitation before and after compaction. This parameter was demonstrated during the essays in true size by the measurement of the particle size analysis and MBV before and after compacting by means of different compaction energies. The results show (Table 4) that the maximum particle size evolution of 18.39%(explained by the presence of friable limestone and marly rocks characterized by low mechanical resistance) was recorded with eight passes, almost identical evolution of 18.23% was recorded in the case of four passes which concerns the evolution towards the fine fraction, this explains that the production of the fine elements is stopped under the effect of compaction energy of four passes, it was also noted that MBV values increased slightly with increasing compaction energy; even with these evolutions, the material always keeps the same classification family after compaction (C_1B_5) according to the NF-P11-300 [24] standard.

Table 4. Granulometric and Methylene Blue value (MBV) evolution according to the compaction energy.

State of Material	Before Compaction	After Compaction						
Compaction Energy		2 Passes	Evolution (%)	4 Passes	Evolution (%)	8 Passes	Evolution (%)
<0.08 mm (wt. %)	19.20	21.00	9. 38	22. 70	18.23	22. 73	18. 39
<2 mm (wt. %)	37.60	40.20	1.60	41.00	9.04	41.10	9.31
<20 mm (wt. %)	63.80	68.00	6.58	71.00	11. 29	71.20	11.60
<50 mm (wt. %)	89.20	95.00	6. 50	96.00	7. 62	96. 10	7.74
<2mm (0/50 mm) (wt. %)	42.15	42.32	0.39	42.71	1.32	42.77	1.46
<0.08 mm (0/50 mm) (wt. %)	21.52	22.11	2.70	23.65	9.85	23.65	9.89
MBV (g/100g)	0.57	0.61	7.02	0.66	15.79	0.67	17.54

The in situ tests have shown that with a very important compaction energy of 8 passes, the process of alteration of the blocks has been almost stopped (granulometric evolution less than 0.9% for fine elements and less than 3% for other diameters found by adding four additional passes of the compactor to get to 8 passes). The results of the dry density according to the compacting energy at the surface and at the bottom of the compacted layer have been summarized in the Table 5. Embankment dry density was examined as a function of roller passes.

Table 5. The compaction rate according to the compaction energy.

Layer		First Layer			Second Layer			Third Layer		
Compaction Energy		2 Passes	4 Passes	8 Passes	2 Passes	4 Passes	8 Passes	2 Passes	4 Passes	8 Passes
reference dry density (kN/m^3)						19.5				
Surface	dry density (kN/m^3)	18.2	19.4	19.0	18.4	19.4	19.1	18.2	19.5	19.1
	compaction rate (%)	93	99	97	94	99	98	93	100	98
Bottom	dry density (kN/m^3)	17.4	18.2	18.1	17.5	18.0	18.0	17.3	18.3	17.8
	compaction rate (%)	90	96	95	91	96	94	90	95	96

The results showed that the maximum dry density corresponding to 19.5 kN/m^3 was recorded for the compaction energy of four passes, the reference dry density is therefore taken equal to 19.5kN/m^3.The calculation of the compaction rates (at the surface and at the bottom of the layer) makes it possible to show that the application of a compaction energy of two passes does not satisfy the required compaction levels for embankments contrary to four and eight passes (Table 5). After application of each compaction energy, the bearing capacity test was carried out just after the compaction of the third layer by ensuring the average water moisture of the material at the time and after compaction, this has been verified by sampling during the entire duration of the measurements by realizing a sounding through the three layers (Table 6).

Table 6. Results of lift tests.

Compaction Energy			2 Passes	4 Passes	8 Passes
Plate test (average of six points)	standard deviation	%	2.23	2.52	2.68
	EV_1	MPa	45.70	57.30	57.10
	EV_2	MPa	79.80	91.20	89.40
	K (EV_2/EV_1)	-	1.75	1.59	1.57

With k (EV_2 / EV_1) < 2 (which makes it possible to appreciate the quality of the compaction) and an average EV_2 module > 80 MPa, material I2 has very satisfactory lift results. According to the LCPC-Setra guide [41], these materials, which are also sensitive to water, can therefore, be classified as Top part of the earthworks (PST3) from a class (AR2) formation if the constructive drainage arrangements make it possible to evacuate the water and prevent its infiltration.

The aforementioned criteria of grain size evolution, densities and bearing capacity justify the choice to be limited only to the compaction energy of four passes for the construction of embankments with PMWR. this retained energy, which remains more important than that required by the LCPC-Setra guide (limited to only three roller passes and for the same compaction parameters), allows to have a maximum fractionation especially for the marly rocks recognized by their evolving behavior and therefore avoid having two fractions with clearly differentiated granulometry (large particles and very fine fraction), this has been demonstrated by the realization of in situ trenches which show that the compacted material is coherent (the fines perfectly fill the voids between the blocks), resistant, and has a homogeneous appearance (a reduced standard deviation found during densities and lift measurements over the entire thickness of the compacted layer).

In view of the aforementioned results, the optimal compaction conditions which make it possible to obtain the compaction level required for embankments (by humidification), to ensure the minimum bearing capacity for the embankment materials and to avoid possible disorders due to the phenomenon of grain size evolution under the effect of mechanical stresses are the following (Table 7).

Table 7. Optimal compaction conditions of I2 material.

Moisture Content (wt. %)	Compactor Class	Compactor Speed (km/h)	Compaction Energy	Thickness (m)
Average (0.9 à 1.1) w_{opn}	V4 (vibratory compactor roller)	4	4 passes	0.30

3.3. Risks Factors Evaluation

The results of the oedometric test showed that the material I2 has An average oedometric modulus of 10,045 kPa, a low compressibility index of 0.125 and a very low swelling index of 0.04, considering that the contracting regulations in Morocco often require settlements of less than 10 cm in 25 years on ordinary road embankments, an evaluation study of embankment settlements (in case of construction with I2 materials) was carried out to determine the maximum height beyond which stability will be questioned, the results make it possible to conclude that this material can be used for embankments up to a height of 15 m respecting a minimum rate of compaction of 95% of γdr and a water content close to w_{opn} without any significant risk of instability. With an organic matter content well below 3% threshold required by the NF P11-300 standard, these releases are therefore far from the category of organic materials, with the availability of deposits, the passage of this materials from the status of waste to an alternative material can therefore be pronounced. This leads back to identifying and evaluating the possible risk factors resulting from this study (Table 8).

Table 8. Risk assessment.

Characteristics of the Material	Risk	Proposed Remedies
Limited mechanical strength (presence of clay)	• Granulometric evolution towards the fine fraction (creation of fine elements under the effect of mechanical stresses which can influence the water resistance) • settlement due to the collapse of fine fraction • swelling	respect the optimum conditions of use (in situ tests)
Presence of rock of different petrographic origin (heterogeneity)	• Settlement for high embankments (>15m) • (difference in drainage and water permeability in the embankment)	• Good identification of deposits • Representative sampling • Rigorous control of traceability and homogenization of storage process • The use is limited to embankments with heights less than 15 m (otherwise, special construction arrangements for stability, embankment base, circulation and drainage of water, variation of permeability will be required)

4. Conclusions

This study is the first of its kind consisting of a physicochemical, mineralogical, environmental and geotechnical characterization of phosphate mine waste rocks. The main conclusions from the laboratory and in situ trial tests to assess the potential use of these materials for road construction are the following:

- The mechanical behavior of these materials depends essentially on their flintstone and clay content.
- The chemical and mineralogical composition and leaching tests on PMWR suggests that they are chemically inert.
- The in situ full trial testing has defined the optimal compaction condition for the use of PMWR in ordinary embankment construction (used in a wet way). It consists of a compaction energy of four passes, a speed of a V4 vibratory roller compactor of 4 km/h and a thickness of the compacted layers of 30 cm.
- Embankments up to 15 m height can be built with PMWR without any significant physical instability risks. The respect of the constructive provisions is necessary.

Considering the results of the leaching tests, the organic content and geotechnical properties, the PMWR can be assimilated to the category of conventional natural aggregates. The use of these materials will have a very important impact on the preservation of the use of natural resources (avoiding the use of borrowing materials) and the recycling of PMWR.

Even with an important level of heterogeneity linked to several scales: the extraction mode, the storage method, and the petrography of the original rocks. It may be recommended that PMWRs be considered as alternative aggregates to be sorted according to a pre-defined zoning map in order to simplify their reuse in civil engineering. The Moroccan guide for road (GMTR) should be updated to allow PMWR to be classified as natural aggregates.

Author Contributions: M.A. conducted all the physical and geotechnical characterization tests while Y.T. conducted all the chemical, mineralogical and environmental characterizations. The full-scale trial was conducted and supervised by M.A. and Y.T. The interpretation of results and paper writing were done by M.A. and Y.T. under the supervision of R.H., A.K. and M.B.

Funding: This research was funded by OCP-SA and CNRST Morocco project APHOS, MAT-HAK-01/2017.

Acknowledgments: The authors thank OCP-SA and CNRST Morocco for their support in the framework of the project APHOS, MAT-HAK-01/2017. The authors greatly acknowledge the OCP-SA collaborators for the great help concerning the sampling, and full-scale trial tests conducting. The authors are also grateful to the staff of Mohammed VI Polytechnic University for their valuable contribution to the solid samples characterizations.

Conflicts of Interest: The authors declare no conflict of interest.

Acronyms and Abbreviations

Corg	Organic carbon content
LA	Los Angeles abrasion value
MD	Micro Deval value
TCLP	Toxicity characteristic leaching procedure
XRD	X-ray diffraction
PMWR	Phosphate mine waste rocks
w_{opn}	Optimum water content of the standard Proctor test
CBR	California bearing ratio
MBV	Methylene blue value
ρs	specific (particle) density
γdr	reference dry density
OCDE	Organisation de coopération et de développement économiques

References

1. OCDE. *Utilisation des Déchets et Sous-Produits en Technique Routière Système d'archivage DSpace/Manakin*; OCDE: Paris, France, 1977.
2. Sayagh, S. *Approche Multicritère de l'utilisation des Matériaux Alternatifs Dans Les Chaussées*; Ecole des Ponts ParisTech: Marne-la-Vallée, France, 2007.
3. Lidelöw, S.; Mácsik, J.; Carabante, I.; Kumpiene, J. Leaching behaviour of copper slag, construction and demolition waste and crushed rock used in a full-scale road construction. *J. Environ. Manag.* **2017**, *204*, 695–703. [CrossRef] [PubMed]
4. Anupam, A.K.; Kumar, P.; Ransinchung, G. Use of various agricultural and industrial waste materials in road construction. *Procedia-Soc. Behav. Sci.* **2013**, *104*, 264–273. [CrossRef]
5. Toraldo, E.; Saponaro, S.; Careghini, A.; Mariani, E. Use of stabilized bottom ash for bound layers of road pavements. *J. Environ. Manag.* **2013**, *121*, 117–123. [CrossRef] [PubMed]
6. Lynn, C.J.; Ghataora, G.S.; OBE, R.K.D. Municipal incinerated bottom ash (MIBA) characteristics and potential for use in road pavements. *Int. J. Pavement Res. Technol.* **2017**, *10*, 185–201. [CrossRef]
7. Vizcarra, G.O.C.; Casagrande, M.D.T.; da Motta, L.M.G. Applicability of municipal solid waste incineration ash on base layers of pavements. *J. Mater. Civ. Eng.* **2013**, *26*, 06014005. [CrossRef]
8. Arulrajah, A.; Piratheepan, J.; Disfani, M.M.; Bo, M.W. Geotechnical and geoenvironmental properties of recycled construction and demolition materials in pavement subbase applications. *J. Mater. Civ. Eng.* **2012**, *25*, 1077–1088. [CrossRef]
9. Cardoso, R.; Silva, R.V.; de Brito, J.; Dhir, R. Use of recycled aggregates from construction and demolition waste in geotechnical applications: A literature review. *Waste Manag.* **2016**, *49*, 131–145. [CrossRef] [PubMed]
10. Shen, W.; Zhou, M.; Ma, W.; Hu, J.; Cai, Z. Investigation on the application of steel slag–fly ash–phosphogypsum solidified material as road base material. *J. Hazard. Mater.* **2009**, *164*, 99–104. [CrossRef] [PubMed]

11. Cuadri, A.; Navarro, F.; García-Morales, M.; Bolívar, J. Valorization of phosphogypsum waste as asphaltic bitumen modifier. *J. Hazard. Mater.* **2014**, *279*, 11–16. [CrossRef] [PubMed]
12. Sorlini, S.; Sanzeni, A.; Rondi, L. Reuse of steel slag in bituminous paving mixtures. *J. Hazard. Mater.* **2012**, *209*, 84–91. [CrossRef] [PubMed]
13. Eskioglou, P.; Oikonomou, N. Protection of environment by the use of fly ash in road construction. *Glob. Nest J.* **2008**, *10*, 108–113.
14. Dubois, V.; Abriak, N.E.; Zentar, R.; Ballivy, G. The use of marine sediments as a pavement base material. *Waste Manag.* **2009**, *29*, 774–782. [CrossRef] [PubMed]
15. Wang, D.X.; Abriak, N.E.; Zentar, R.; Xu, W. Solidification/stabilization of dredged marine sediments for road construction. *Environ. Technol.* **2012**, *33*, 95–101. [CrossRef]
16. Zentar, R.; Dubois, V.; Abriak, N.E. Mechanical behaviour and environmental impacts of a test road built with marine dredged sediments. *Resour. Conserv. Recycl.* **2008**, *52*, 947–954. [CrossRef]
17. Miraoui, M.; Zentar, R.; Abriak, N.-E. Road material basis in dredged sediment and basic oxygen furnace steel slag. *Constr. Build. Mater.* **2012**, *30*, 309–319. [CrossRef]
18. Hakkou, R.; Benzaazoua, M.; Bussière, B. Laboratory evaluation of the use of alkaline phosphate wastes for the control of acidic mine drainage. *Mine Water Environ.* **2009**, *28*, 206. [CrossRef]
19. Ouakibi, O.; Loqman, S.; Hakkou, R.; Benzaazoua, M. The potential use of phosphatic limestone wastes in the passive treatment of AMD: A laboratory study. *Mine Water Environ.* **2013**, *32*, 266–277. [CrossRef]
20. Ouakibi, O.; Hakkou, R.; Benzaazoua, M. Phosphate carbonated wastes used as drains for acidic mine drainage passive treatment. *Procedia Eng.* **2014**, *83*, 407–414. [CrossRef]
21. Bossé, B.; Bussière, B.; Hakkou, R.; Maqsoud, A.; Benzaazoua, M. Assessment of phosphate limestone wastes as a component of a store-and-release cover in a semiarid climate. *Mine Water Environ.* **2013**, *32*, 152–167. [CrossRef]
22. GMTR. *Direction des Routes et de la Circulation Routière. 2002. Guide Marocain pour les Terrassements Routier, Fascicule 1*; Royaume du Maroc; Ministère de l'Equipement: Rabat, Morocco, 2002; 106p.
23. Ahmed, A.; Abouzeid, A. Potential use of phosphate wastes as aggregates in road construction. *J. Eng. Sci* **2009**, *37*, 413–422.
24. NF-P11-300. *Exécution des Terrassements—Classification des Matériaux Utilisables dans la Construction des Remblais et des Couches de Forme d'infrastructures Routières*; Association Française de Normalisation: Paris, France, 1992.
25. EPA-1311. *SW-846 Test Method 1311: Toxicity Characteristic Leaching Procedure*; EPA: Washington, DC, USA, 1992.
26. US-EPA. *Hazardous Waste Characteristics. A User-Friendly Reference Document*; EPA: Washington, DC, USA, 2009.
27. ISO-14235. *Soil Quality—Determination of Organic Carbon by Sulfochromic Oxidation*; ISO: Geneva, Switzerland, 1998.
28. NF-P94-048. *Soil: Investigation and Testing—Determination of the Carbonate Content—Calcimeter Method—Sols: Reconnaissance et essais*; Association Française de Normalisation: Paris, France, 1996.
29. NF-P94-050. *Soils: Investigation and Testing. Determination of Moisture Content. Oven Drying Method*; Association Française de Normalisation: Paris, France, 1995.
30. NF-P94-051. *Soil: Inverstigation and Testing. Determination of Atterberg's Limits. Liquid Limit Test Using Cassagrande Apparatus. Plastic Limit Test on Rolled Thread—Sols: Reconnaissance et Essais*; Association Française de Normalisation: Paris, France, 1993.
31. NF-P94-052-1. *Soil: Investigation and Testing. Atterberg Limit Determination. Part 1: Liquid Limit. Cone Penetrometer Method—Sols: Reconnaissance et essais*; Association Française de Normalisation: Paris, France, 1995.
32. NF-P94-056. *Soil: Investigation and Testing. Granulometric Analysis. Dry Sieving Method after Washing—Sols: Reconnaissance et essais*; Association Française de Normalisation: Paris, France, 1996.
33. NF-P94-068. *Soils: Investigation and Testing. Measuring of the Methylene Blue Adsorption Capacity of à Rocky Soil. Determination of the Methylene Blue of à Soil by Means of the Stain test—Sols: Reconnaissance et essais*; Association Française de Normalisation: Paris, France, 1998.
34. NF-P94-093. *Soils: Investigation and Testing—Determination of the Compaction Reference Values of a Soil Type—Standard Proctor Test—Modified Proctor Test—Sols: Reconnaissance et essais—Détermination des références de compactage d'un matériau—Essai Proctor Normal—Essai Proctor modifié*; Association Française de Normalisation: Paris, France, 2014.

35. P18-573. *Aggregates. Los Angeles Test—Granulate. los Angeles Pruefung*; Association Française de Normalisation: Paris, France, 1990.
36. P18-572. *Aggregates. Micro-Deval Attribution Test*; Association Française de Normalisation: Paris, France, 1990.
37. NF-P94-066. *Soils: Investigation and Tests. Fragmentability Coefficient of Rocky Material—Sols: Reconnaissance et essais*; Association Française de Normalisation: Paris, France, 1992.
38. NF-P94-067. *Soils: Investigation and Tests. Degradability Coefficient of ROCKY material—Sols: Reconnaissance et essais*; Association Française de Normalisation: Paris, France, 1992.
39. NF-P94-078. *Soils: Investigation and Tests. CBR after Immersion. Immediate CBR. Immediate Bearing Ratio. Measurement on Sample Compacted in CBR Mould—Sols: Reconnaissance et essais*; Association Française de Normalisation: Paris, France, 1997.
40. NF-P94-071-1. *Soil Investigation and Testing. Direct Shear Test with Shearbox Apparatus. Part 1: Direct Shear—Sols: Reconnaissance et essais*; Association Française de Normalisation: Paris, France, 1994.
41. XP-P94-090-1. *Soil: Investigation and Testing. Oedometric Test. Part 1: Compressibility Test on Quasi Satured Fine Grained Soil with Loading in Increments*; Association Française de Normalisation: Paris, France, 1997.
42. NF-P94-061-2. *Soils: Investigation and Testing. Determination of Density of Materials on Site. Part 2: Membrane Densitometer Method—Sols: Reconnaissance et Essais*; Association Française de Normalisation: Paris, France, 1996.
43. NF-P98-736. *Road Construction and Maintenance Equipment. Compactors. Classification—Matériel de Construction et d'entretien des Routes*; Association Française de Normalisation: Paris, France, 1992.

© 2019 by the authors. Licensee MDPI, Basel, Switzerland. This article is an open access article distributed under the terms and conditions of the Creative Commons Attribution (CC BY) license (http://creativecommons.org/licenses/by/4.0/).

Article

Utilisation of Water-Washing Pre-Treated Phosphogypsum for Cemented Paste Backfill

Yikai Liu [1], Qinli Zhang [1], Qiusong Chen [1,*], Chongchong Qi [2], Zhu Su [1] and Zhaodong Huang [1]

1. School of Resources and Safety Engineering, Central South University, Changsha 410083, China; yikai1995@foxmail.com (Y.L.); zhangqinlicn@126.com (Q.Z.); suzhu16@163.com (Z.S.); huangzhaodong1995@foxmail.com (Z.H.)
2. School of Civil, Environmental and Mining Engineering, The University of Western Australia, Crawley 6009, Australia; chongchong.qi@gmail.com
* Correspondence: qiusong.chen@csu.edu.cn; Tel.: +86-151-1627-9873

Received: 22 January 2019; Accepted: 9 March 2019; Published: 12 March 2019

Abstract: Recycling phosphogypsum (PG) for cemented paste backfill (CPB) has been widely used at phosphate mines in China. However, the impurities in PG prolong the setting time and reduce the uniaxial compressive strength (UCS), limiting the engineering application of PG. This paper aims to investigate the feasibility of treated PG (TPG) washed repeatedly using deionised water (DW) for CPB. A water-washing pre-experiment was first conducted to find the proportion with the least DW demand and the effects of water-washing on ordinary PG (OPG). Then, based on the PG:DW ratio obtained from the pre-experiment, the properties of the OPG-based CPB (OCPB) and TPG-based CPB (TCPB) were tested using slump tests, UCS tests, and microstructural analysis. The results show that (1) after 11 water-washings at the PG:DW ratio of 1:1.75, the pH of the supernatant (pH = 6.328) meets the requirements of Chinese standard GB 8978-1996. (2) Water-washing improves the particle gradation quality of PG and removes the soluble impurities adsorbed at the surface of PG crystals. (3) The initial slump values of TCPB are 0.19–1.15 cm higher than that of OCPB, furthermore, the diffusivity values of TCPB are better than the performance of OCPB, with 0.61–1.68 cm of superiority. (4) The UCS values of TCPB are up to 0.838 MPa, 1.953 MPa, and 2.531 MPa, after curing for 7, 14, and 28 days. These are 0.283 MPa, 0.823 MPa, and 0.881 MPa higher than that of OCPB, respectively. It can be concluded that water-washing pre-treatment greatly improves the workability and mechanical property of PG-based CPB. These results are of great value for creating a reliable and environmentally superior alternative for the recycling of PG and for safer mining production.

Keywords: phosphogypsum; purification; water washing; waste recycle; cemented paste backfill

1. Introduction

Phosphogypsum (PG) is the main solid waste from the production of ammonium phosphate and phosphoric acid raw materials. An industrial plant of phosphate fertilisers produces approximately 4–6 tonnes of PG for every ton of phosphoric acid produced. The total amount of PG produced up to 2006 was estimated to be around 6 billion tonnes. The annual production of PG worldwide was estimated to be around 160 million tonnes. The production of PG is increasing worldwide and could reach 200–250 million tonnes within the next decade or two [1]. Normally, PG is mostly composed of calcium sulphate dihydrate ($CaSO_4 \cdot 2H_2O$) containing some impurities, such as phosphoric acid (H_3PO_4), hydrofluoric acid (HF), heavy metals (Sr, Ba, Cu, Cd, etc.), and radioactive-elements (^{226}Ra, ^{238}U) [1,2]. Historically, PG was disposed of on the Earth's surface, which not only wasted a lot of resources, but also caused serious pollution of the atmosphere, soil, and water near the burial locations [3–7].

The effective utilisation of PG as an industrial waste is closely related to the coordinated development of natural resources, environment, and economy. However, because of the physical and chemical characterisations of PG, strong acidity (pH < 3) and high moisture content, only about 15% of PG has been reused in different fields such as soil stabilisation amendments, agricultural fertilisers, as set controller in cement manufacture, and in building materials [8–11]. Recently, researchers [12–14] have applied PG in cemented paste backfill (CPB) for use in underground mine stopes, which has been approved as an economic, safe, and environmentally friendly technology. Specifically, CPB is an engineered material created by the combination of different types of binders, tailings having different physical characteristics and the water [15–19]. In addition, owing to its high content of $CaSO_4 \cdot 2H_2O$, which produces insoluble $Ca_3(PO_4)_2$ and hinders the hydration process, ordinary PG (OPG) is a typical cement retarder. Moreover, the impurities in OPG, such as HF and H_3PO_4, also prolong the setting time and reduce the uniaxial compressive strength (UCS) of CPB. In this case, researchers have suggested different methods to ensure the backfill quality of PG used as an alternative raw material for application in CPB [16–24]. These include, for instance, substantially increasing the proportion of ordinary Portland cement (OPC) in mixtures or adding other tailings, such as fly ash, waste lime and zeolite, as the hydration activators. Romero-Hermida et al. [25] investigated the rheological properties and microstructures of special lime putty prepared from PG. These methods decrease the setting time and improve the mechanical property of PG-based CPB. However, the cost is greatly increased, which substantially exceeds the resource capacity of (profitable) mining enterprises and limits the application of PG.

Therefore, many researchers tried to remove or at least reduce the impurities in PG to guarantee its safe use in the construction field. Taher [26] conducted a comparative study of PG with thermal treatments at different temperatures in an attempt to purify PG and improve its performance as a tailing. The results showed that thermal treatments can effectively improve the hydraulic properties of Portland slag cement and the best hydraulic properties of Portland slag cement occurred when using PG thermally treated at 800 °C. Singh [27] treated PG with aqueous citric acid solution, intending to convert phosphatic and fluoride impurities into water-removable citrates, aluminates and ferrates. The results of this study illustrated that the purified PG had less content of phosphates, fluorides and organic matter compared with the untreated material. Mun et al. [28] used 0.5% milk of lime at 20 °C to wash PG for 5 min (the ratio of PG:milk was 14%), after neutralisation treatment, the PG was dried at 80 °C. In this study, the pre-treated PG was not only considered as an effective activator, but also an impactful binder to granulated blast-furnace slag. Potgieter et al. [29] illustrated that treatment with ammonium hydroxide or sulphuric acid was effective in reducing set retardation of PG-based CPB. The above pre-treatment methods can remove the impurities in PG to different degrees, but the properties of the prepared mixture are different due to the different mechanisms of the pre-treatment methods. Meanwhile these methods are too complicated in the field for industrial practice. In this research, thus, a pre-treatment method, water-washing, was proposed to remove impurities from PG.

The pre-treatment of adding water and separating supernatant and precipitate by centrifuge is widely used as an effective and simple method of impurity removal in various industries. Kamimura et al. [30] studied an efficient purification method for washing recovered fibre-reinforced plastic monomer with water, which improved the hardness of recycled plastics and made them as hard as polyester made from virgin materials. Slavinskaya [31] used water to remove organic and mineral impurities at an industrial ion-exchange installation. Shuang et al. [32] investigated the possibility of using deionised water (DW) to remove chloride ions for inexpensive ultra-pure $TiCl_4$. Therefore, as a simple and mild pre-treatment method for purification, water-washing has potential application for PG-based CPB, especially considering that the $CaSO_4 \cdot 2H_2O$ in PG is slightly soluble in water and most of the impurities, such as H_3PO_4, HF, and ammonium salt, are also soluble in water [33]. Cárdenas-Escudero et al. [34] demonstrated that the alkaline soda solution can effectively dissolve PG and form the portlandite precipitation.

In the present study, investigations were carried out to study the chemical and physical optimisations of water-washing pre-treatment to PG for using as the raw materials in PG-based CPB. A water-washing pre-experiment was firstly conducted to optimise the proportion of PG:DW recording the total nitrogen (TN), chloride ion concentration (Cl), fluoride concentration (F), total phosphorus (TP), conductance (Cond) and pH parameters of the supernatant liquid in each washing. Then, scanning electron microscopy (SEM), X-ray diffraction (XRD) analysis, X-ray fluorescence (XRF) analysis, and particle size tests were conducted. Secondly, the properties of OPG-based CPB (OCPB) and treated PG (TPG)-based CPB (TCPB) were determined using slump tests, diffusivity tests, UCS tests, and microstructural analysis. The flow chart of this work is shown in Figure 1.

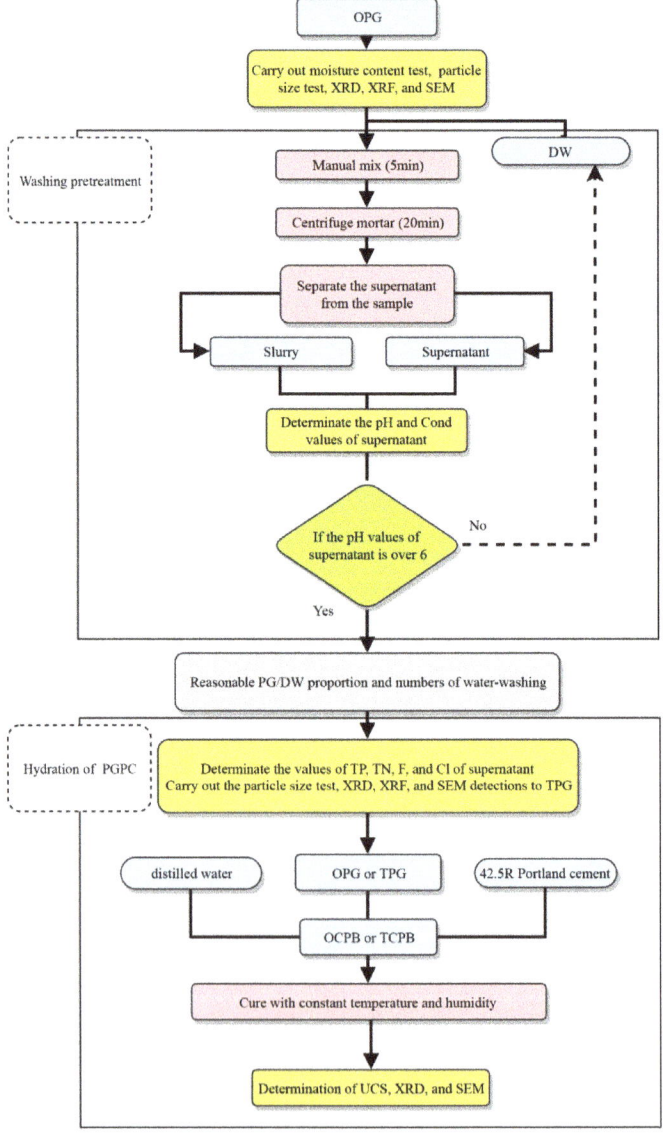

Figure 1. Flow chart of the experiments.

2. Materials and Methods

2.1. Raw Materials and Water-Washing Pre-Treatment

OPC 42.5R (similar to ASTM C150 Type I cement [35]) and OPG with moisture content of 18.26%, natural density (wet) of 1.62 g/cm^3, and dry density of 0.86 g/cm^3 (heated for 24 h, 95 ± 5 °C), which was obtained from a phosphoric acid factory named Western Chemical Co. Ltd., plant (Yichang, China), were used in this work. The physical properties and chemical compositions of OPC and OPG are shown in Tables 1 and 2, respectively, and the particle size distribution of OPG and OPC are shown in Figure 2, the data were provided using a laser particle size tester (Master Sizer 2000, Malvern Instruments Ltd., Malvern, UK). As stated above, the main components of OPG are SO$_3$ (41.504%) and CaO (44.897%), with a small amount of P$_2$O$_5$ (1.224%) and F (0.817%). The particle size of OPG ranges from 0.710 μm to 255.53 μm and the mean particle size (D_{50}) is 58.329 μm. In addition, the water used in the washing pre-treatment was DW (produced by Solar-bio Co. Ltd., Beijing, China) to avoid the effects of ions, while the water used in the hydration of OCPB and TCPB was ordinary distilled water.

Table 1. Chemical compositions of ordinary Portland cement (OPC) and ordinary phosphogypsum (OPG).

Chemical Composition (%)	OPC	OPG
Na$_2$O	0.090	0.150
MgO	1.840	0.111
Al$_2$O$_3$	4.420	0.782
SiO$_2$	17.94	5.730
P$_2$O$_5$	0.215	1.224
SO$_3$	3.400	41.504
Cl	0.022	-
K$_2$O	0.671	0.674
CaO	60.850	44.897
TiO$_2$	0.307	0.648
V$_2$O$_5$	0.040	-
Cr$_2$O$_3$	0.025	-
MnO	0.265	-
Fe$_2$O$_3$	3.814	3.208
CuO	0.015	-
ZnO	0.043	-
Rb$_2$O	0.003	-
SrO	0.067	0.142
ZrO$_2$	0.010	-
BaO	0.034	0.113
F	-	0.817
Loss	5.929	-

Table 2. Physical properties of OPC and OPG.

Physical Properties	OPC	OPG
SSA (m^2/g)	1.690	0.324
$D_{[4,3]}$ (μm)	24.449	69.935
D_{10} (μm)	1.426	13.176
D_{50} (μm)	14.641	58.329
D_{90} (μm)	64.392	146.589
$D_{d<20\,\mu m}$ (%)	57.84	81.78
Moisture content (%)	-	18.26
Natural density (g/cm^3)	2.94	1.62
Dry density (g/cm^3)	-	0.86

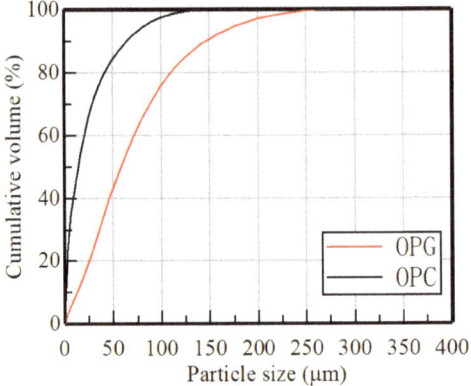

Figure 2. Particle size distribution of OPG and OPC.

In this study, eight PG:DW ratios were used in the pretreatment, ranging from 0.4 to 2.0. The PG measure was wet weighed before mixing with DW of certain qualities according to Table 3 and then artificially stirred for 5 min in a steel blender (TD5M low-speed desktop centrifuge, produced by Shanghai Lu Xiangyi Centrifuge Instrument Co. Ltd., Shanghai, China), subsequently, the uniformly mixed materials were centrifuged for 20 min at 300 r/min. According to the Chinese standard of surface water environmental quality [36], the fifth-class surface water (suitable for agricultural water and water for general landscapes, etc.) is guaranteed because the content of TP, TN, F are less than 0.4, 2.0, 1.5 mg/L. Besides, in central drinking water surface, the content of Cl in the discharged water should be less than 250 mg/L. Therefore, after centrifugation, the supernatant was taken out to determine the pH and Cond using a multiparameter analyser (produced by INESA, DZS-708L multiparameter controller), ionic electrodes (produced by INESA, BestLab water pH composite electrode), and a conductance electrode (produced by INESA, DJS-10C conductance electrode). The washing process was repeated until the pH of supernatant fully met the requirements of Chinese standard (pH > 6) [37]. The measured supernatant was collected in plastic sealed cups and kept at 25 ± 2 °C using a thermostat. The content of F and Cl were tested using the same multiparameter analyser and other ionic electrodes produced by INESA (PF-202-C fluorine ion composite electrode, PCL-1-01 chloride ion electrode). The values of TP and TN were determined using a water quality measurement instrument (5B-6C-V8, produced by Lianhua Science and Technology Co. Ltd., Beijing, China). The electrodes were washed with DW after each measurement.

Table 3. Mix proportions of OPG samples.

No.	Weight of OPG		DW (g)	PG:DW	Number of Washes	Total DW (g)
	Wet (g)	Dry (g)				
A	1000.00	817.40	226.10	1:0.50	79	17,861.90
B	1000.00	817.40	430.45	1:0.75	46	19,800.70
C	1000.00	817.40	634.80	1:1.00	37	23,487.60
D	1000.00	817.40	839.15	1:1.25	28	23,496.20
E	1000.00	817.40	1043.50	1:1.50	19	19,826.50
F	1000.00	817.40	1247.85	1:1.75	11	13,726.35
G	1000.00	817.40	1452.20	1:2.00	10	14,522.00
H	1000.00	817.40	1860.90	1:2.50	8	14,887.20

2.2. Mix Proportions and Cemented Paste Backfill (CPB) Preparation Process

As shown in Table 4, there were 12 different mix-proportions of PG-based CPB with all materials being calculated by mass percent. In accordance with the above proportions, all solid materials were weighed and mixed in the planetary mixer for 30 s before adding water. In addition, when weighing the raw materials, PG was considered as dry mass by subtracting the water content. After that, raw materials were stirred homogeneously for 5 min and the premixed mixture was casted into plastic cylindrical moulds [38] (diameter of 5 cm and height of 10 cm). After 24 h of hydration, specimens were taken out of the moulds and moved into a standard chamber with constant temperature and humidity (25 ± 2 °C and 90 ± 5%, respectively).

Table 4. Mix proportions of PG-based cemented paste backfill (CPB).

No.	OPC	OPG (Dry)	TPG (Dry)	Water	Water/Solid	Water/Cement
1-OPG	8.57%	62.92%	-	28.51%	0.67	3.33
2-OPG	8.86%	65.01%	-	26.13%	0.61	2.95
3-OPG	6.67%	65.25%	-	28.09%	0.67	4.21
4-OPG	6.89%	67.42%	-	25.69%	0.61	3.73
5-OPG	5.45%	66.73%	-	27.82%	0.67	5.10
6-OPG	5.64%	68.95%	-	25.41%	0.61	4.51
1-TPG	8.57%	-	62.92%	28.51%	0.67	3.33
2-TPG	8.86%	-	65.01%	26.13%	0.61	2.95
3-TPG	6.67%	-	65.25%	28.09%	0.67	4.21
4-TPG	6.89%	-	67.42%	25.69%	0.61	3.73
5-TPG	5.45%	-	66.73%	27.82%	0.67	5.10
6-TPG	5.64%	-	68.95%	25.41%	0.61	4.51

2.3. Test Methods

2.3.1. Test of X-ray Fluorescence (XRF)

Elemental analysis was carried out on a Bruker S4 Pioneer XRF analyser (Bruker, Billerica, MA, USA) and the samples were processed by the conventional pressing plate method. The samples were first ground into powder by a grinder, after which they were heated in an oven at 45 ± 5 °C for 24 h. Then, 2 g of the processed sample was accurately weighed and blended well with 2 g of boric acid in an agate mixture. Next, the mixed powder was loaded into a die and compacted (60 s at 40 t pressure) into a circular plate with a diameter of 32 mm and an outer diameter of 40 mm using the boric acid to pad the bottom edge. Samples were labelled and preserved in a dry and pollution-free environment after pressing and then analysed using full quantitative analysis without standard samples.

2.3.2. Test of Abrams Cone

Much of the research was to determine if the CPB mixture with a slump value over 170 mm could be successfully transported underground by backfill pumps [39–41]. Therefore, in this study, trial slump tests [9] were conducted to make sure that the slump values of the OCPB and TCPB mixtures were greater than 170 mm. The preparation of materials and the test process were in accordance with the standard test method for slump [42]. The premixed mixture was casted into a conical mould (the top diameter of 10 cm, the bottom diameter of 20 cm and height of 30 cm) with a plate glass at the bottom. After that, the conical mould was lifted about 10 cm and the previously constrained mixture slurry flowed down freely. The values were measured diagonally and vertically when the mixture stopped flowing and the result was determined by the average of two repeated experiments.

2.3.3. Test of Uniaxial Compressive Strength (UCS)

As a widely used detection mean of the mechanical properties of CPB for exploring an efficient mix design and reaching a higher performance in underground mining structures [43], the standard UCS tests were carried out in this study. For each ratio, there were three corresponding specimens. The UCS tests were carried out according to the ASTM Standard C39 C39M-2014a [44]. After curing for 7, 14, or 28 days, the specimens were subjected to unconfined compressive strength tests using an anti-bending machine (23 MTS Insight, MTS Systems Co. Ltd., Eden Prairie, MN, USA) with 30 kN loading capability at a displacement rate of 0.1 mm/min. Each test was performed in triplicate and the mean value was used for further analysis.

2.3.4. Microstructural Analysis

After the UCS tests, samples were taken from the broken surfaces and then immersed in plastic bottles of acetone solvent in order to terminate the hydration, which bottles were then sealed. Then, superfluous solution was removed using a vacuum filter after 24 h of immersion. The treated samples were studied using a JSM-6490LV scanning electron microscope (JEOL Ltd., Beijing, China). XRD analysis of the samples was performed using an X-ray diffractometer (D8 Advance, Bruker, Karlsruhe, Germany) with CuKα radiation and Ni filter. The XRD patterns were appraised using the references in the PDF-2 data base (PDF-2 International Centre for Diffraction Data, Newtown Square, PA, USA).

3. Results and Discussion

3.1. Effect of Washing Pre-Treatment on Phosphogypsum (PG)

3.1.1. Changes of pH, Conductivity and Values of the Ions in the Supernatant

Table 3 shows the number of water-washing pre-treatments required for different PG:DW proportions. With increase of the PG:DW ratio, the number of washings required increases from 8 to 79. However, the amount of DW required increases to a peak at the PG:DW ratio of 1:1.25 (23,496.20 g); then drops to a bottom at the ratio of 1:1.75 (13,726.35 g). Finally it rises back to 14,887.20 g at the smallest ratio of 1:2.50. Moreover, as is shown in Figure 3a, the Cond values of the different proportions are stable at about 2.00 with increase in the number of washings. After several times of centrifugation, the ions dissolved in water are separated from the PG mixture into the supernatant. This means that stable conductivity is regarded as a sign of the successful washout of soluble impurities in the PG. As regards the changes in pH of the supernatant, Figure 3b shows the variation curves for the different proportions and shows that all the values increase greatly in the first several washings; then the pace of change gradually slows down. This means that in the early stage of the washing process, large quantities of soluble acidic substances in the PG are dissolved in the liquid phase and the conclusion is also consistent with the conductivity of the supernatant. Therefore, taking the amount of DW into consideration, the PG was washed 11 times using DW with PG:DW value of 1:1.75, for use in subsequent experiments.

Figure 4 shows the changes in the dissolved ion concentration in the supernatant with the number of water-washing pre-treatments. At the PG:DW value of 1:1.75, the principal axis demonstrates the content of TN, F and Cl, and the ordinate axis shows the content of TP. After being washed and centrifuged once, the contents of TP, TN and Cl in the supernatant were 5943.78, 23.23 and 264.33 mg/L, respectively. With an increase in the number of washings, the values of TP, TN, and Cl decrease continuously to 10.33, 0.19, and 10.45 mg/L. In contrast, for the first three times, the value of F increases from 30.22 to 52.30 mg/L; then gradually decreases to 12.19 mg/L. The possible reason for the obvious increase in F content is that soluble F mainly occurs in the form of sodium fluoride (NaF) compounds in PG. Meanwhile, as a weak acid, the hydrogen and F ions in the solution form HF only in the highly acidic liquid phase. Therefore, during the initial washing processes, it was easy for the F compounds, which are sparingly soluble in water, to combine with H ions to form HF in the strong acidic liquid phase, thereby causing an increase in the concentration of F in the supernatant. With increase in pH values, the fluoride ions dissolved in the liquid phase were discharged with the supernatant, thus the F-content decreases gradually. Although the content of TN, TP, F, and Cl has eventually decreased after repeated water-washing pre-treatment, the figures are still higher than the standard for direct emission, so the polluted water needs to be put into the sewage treatment plant for further purify as the recycled-water for the concentrator [45].

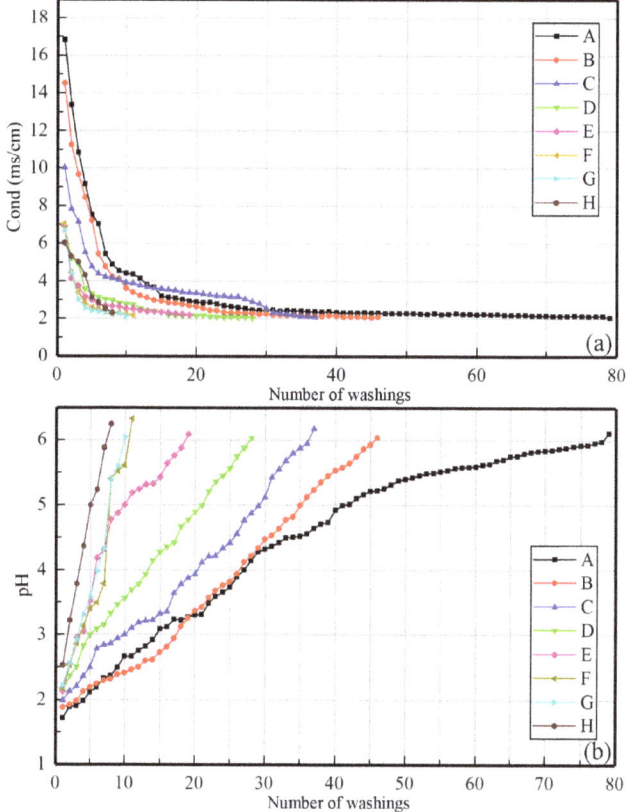

Figure 3. Cond (a) and pH (b) values of the supernatant.

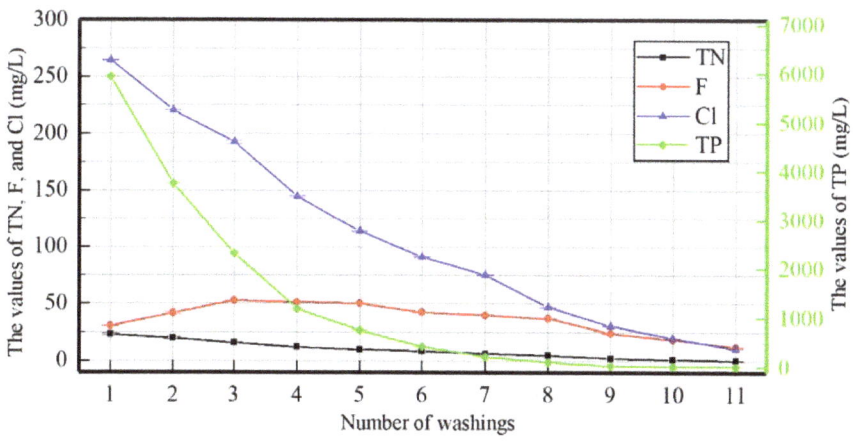

Figure 4. Values of the ion concentration in the supernatant.

3.1.2. Analysis of XRF

As shown in Tables 1 and 5, CaO, SO_3, and SiO_2 are both the main components of OPG and TPG. After the water-washing pre-treatment, the content of CaO and SO_3 in TPG are slightly increased, while the other chemical components decreased relatively. Among them, the content of Fe_2O_3 exhibited the greatest reduction, from 3.208% to 0.663%, which indicates the partial loss of Fe during filtration of the supernatants. Moreover, the content of the impurity P_2O_5 obviously also decreased, from 1.224% to 0.714%. In addition, the impurity MgO was hardly detected in TPG (0.005%), for which the possible reason is that all the MgO exists in PG in the form of water-soluble Mg compounds. The content of the impurities SiO and F in PG showed a modest decrease, while the content of other impurity elements in PG, such as Sr, Ti, and Ba, changed little, suggesting that these were insoluble in water. Therefore, the water-washing pre-treatment can effectively remove some of the soluble compounds in PG, whereas other impurities that are insoluble will likely be difficult to remove.

Table 5. Chemical compositions of TPG.

Chemical Composition (%)	Na_2O	MgO	Al_2O_3	SiO_2	P_2O_5	SO_3	K_2O
TPG	0.065	0.005	0.300	5.194	0.714	45.014	0.298
Chemical Composition (%)	CaO	TiO_2	Fe_2O_3	SrO	BaO	F	Loss
TPG	46.526	0.526	0.663	0.067	0.096	0.532	-

3.1.3. Analysis of Scanning Electron Microscopy (SEM)

Microstructures of OPG and TPG are shown in Figure 5. The crystal grains of OPG and TPG are both rhombic plate particles, which indicates that washing does not change the grain shape of PG. Figure 5a shows that the grain size distribution of OPG was irregular with a lot of small crystals adsorbed on the surface of PG crystal resulting in rough areas of the crystal surface. From Figure 5b, it can be seen that TPG has better surface smoothness with only a small part flocculent crystal but most were columnar and rhombic plates. Apparently, the crystalline form of TPG is better than OPG, with a more complete grain. Besides, although the materials have been washed several times, the particle morphology of TPG is mostly maintained, with a majority of water-soluble impurities removed from the surface of PG.

Figure 5. Scanning electron microscope (SEM) images of the OPG (**a**) and the TPG (**b**).

3.1.4. Analysis of the Particle Size

Table 2, Table 6, and Figure 6 illustrate the particle size of OPG and TPG. The particle size of TPG is obviously smaller, and ranges from 0.0796 µm to 251.785 µm. It is clear that the specific surface area (SSA) and mean-volume diameter ($D_{[4,3]}$) of PG decreased from 0.324 to 0.277 m^2/g and from 69.935 μm to 66.801 μm, respectively, after water-washing pre-treatment. The particle size fitting formulas of OPG and TPG are shown in Equations (1) and (2).

$$C = (1.54224013 \times 10^{-6})x^3 + (-2.97030817 \times 10^{-3})x^2 + 1.03958678x - 0.24821556 \ (R^2 = 0.99881838) \quad (1)$$

$$C = (2.63095018 \times 10^{-6})x^3 + (-3.12349930 \times 10^{-3})x^2 + 1.0208419x - 0.80246342 \ (R^2 = 0.99776654) \quad (2)$$

Table 6. Particle size characters of TPG.

Physical Properties	SSA (m^2/g)	$D_{[4,3]}$ (μm)	D_{10} (μm)	D_{50} (μm)	D_{90} (μm)	$D_{d<20\ \mu m}$ (%)
TPG	0.277	66.801	9.265	57.593	133.816	84.91

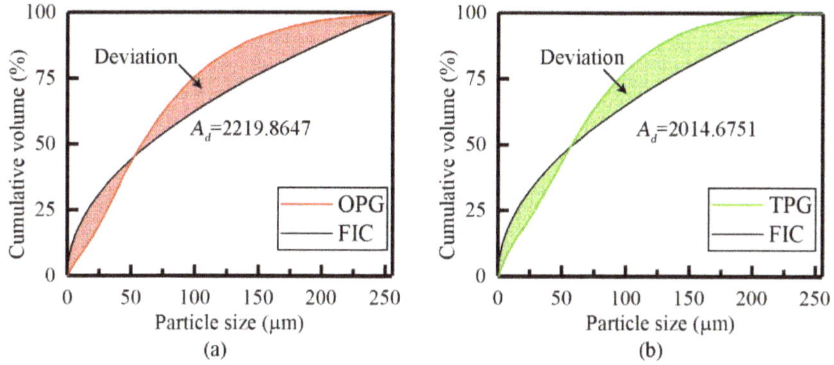

Figure 6. Particle size curve and Fuller ideal curve (FIC) of OPG (a) and TPG (b).

According to the maximum density curve theory proposed by W.B. Fuller et al. [46,47] and Yao [48], it is considered that the closer the tailings gradation curve is to a parabola, the smaller the void between particles and the maximum density. This is mainly used to describe a particle size distribution with continuous gradation. The equation of the Fuller ideal curve (FIC) gradation theory is as follows:

$$P_i = 100 \left(\frac{x_i}{D_{max}} \right)^{0.5} \quad (3)$$

where x_i is the particle size of grade i tailings (mm), D_{max} is the maximum particle size of tailings (mm) and P_i is the passing rate of x_i particles (%). In order to analyse qualitatively the particle size differences of TPG and OPG, the results were calculated using Equation (4) and the curves are shown in Figure 6a,b. The gradation curves of OPG and TPG both obviously deviate from the FIC. Therefore, to study further the influence of water-washing pre-treatment on the gradation of OPG, the area of the tailing's gradation curve deviating from the FIC was used to express the effect; the bigger the deviation area is, the worse the gradation of the raw material is. The area values were calculated using the following equation:

$$A_d = 100 \int_a^b [f(x) - P_i(x)] dx \quad (4)$$

where A_d is the area of tailings gradation curve deviating from ideal Fuller curve, a is the smallest particle size of the raw material and b is the largest particle size of the raw material. The values of A_d are shown in Figure 6, which illustrates that water washing can effectively optimise the particle size gradation with a difference value of 205.1896. Thus, the particle size distribution of PG can be obviously optimised by water-washing pre-treatment, which is closely related to the improvement of UCS values.

3.1.5. Analysis of X-ray Diffraction (XRD)

The results from investigating the purification of PG by washing repeatedly with DW are shown in Figure 7 (XRD patterns of the OPG and the TPG). The mineral phases of the TPG determined from the XRD patterns were $CaSO_4 \cdot 2H_2O$, $CaSO_4 \cdot 0.5H_2O$, $CaSO_4$, and SiO_2. In the case of OPG, $CaSO_4 \cdot 0.5H_2O$, and $CaSO_4$ were undetected. Obviously, the main mineral phases in the OPG and TPG were both $CaSO_4 \cdot 2H_2O$ and SiO_2, However, after being treated, $CaSO_4 \cdot 2H_2O$ in OPG changed to $CaSO_4 \cdot 0.5H_2O$ and $CaSO_4$, resulting in a decline in the content of $CaSO_4 \cdot 2H_2O$ from 98.36% to 83.65%. The reaction equations [49,50], which generally go through the process of dissolution and recrystallisation, are as follows. The equilibrium of Equation (5) is susceptible to the influence of ions in liquid phase and to temperature, which trends to the right side with increasing concentration of H_3PO_4, H_2SO_4, and temperature. Therefore, the presence of phosphate compounds in PG, which form H_3PO_4 in the liquid phase, is one possible factor for the appearance of $CaSO_4 \cdot 0.5H_2O$ in the TPG. However, the crystals of $CaSO_4 \cdot 0.5H_2O$ are metastable and direct dehydration of $CaSO_4 \cdot 2H_2O$ can form $CaSO_4$.

$$CaSO_4 \cdot 2H_2O \rightleftharpoons Ca^{2+} + SO_4^{2-} + 2H_2O \rightleftharpoons CaSO_4 \cdot 0.5H_2O + 1.5H_2O \quad (5)$$

$$CaSO_4 \cdot 2H_2O \rightleftharpoons Ca^{2+} + SO_4^{2-} + 2H_2O \rightleftharpoons CaSO_4 + 2H_2O \quad (6)$$

$$2CaSO_4 \cdot 0.5H_2O \rightleftharpoons 2CaSO_4 + H_2O \quad (7)$$

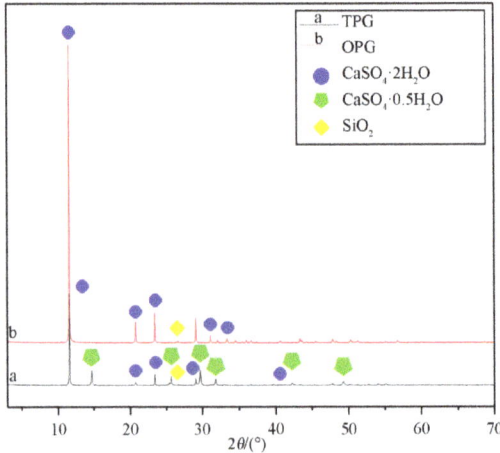

Figure 7. X-ray diffraction (XRD) patterns of OPG and TPG.

3.2. The Workability of PG-Based CPB

In order to analyse the effect of water-washing pre-treatment on the initial and 30 min aging workability of PG-based CPB to obtain relatively qualified workability values, slump and diffusivity tests were carried out. The results of the workability tests of OCPB and TCPB are demonstrated in Figure 8a,b, respectively. The results show that water-washing pre-treatment can obviously optimise the workability of TCPB mixture, furthermore, the slump and diffusivity performances for 30 min of TCPB mixture are still better than that of OCPB regardless of the content of PG, although the rules for increasing the slump and fluidity of the mixtures with different concentrations vary slightly. With an increase of the PG content, the effect from optimisation is more obvious and is at maximum at mix-proportion 6, where the figure for TCPB is 104.5% of OCPB. Meanwhile, the trend is more evident in the mixtures after aging for 30 min, despite the fact that the ratio remains at 104.0%. As regards the values for fluidity of tailings, the initial TCPB mixtures have greater liquidity, compared with the OCPB mixtures, especially at mix-proportion 5, the fluidity of initial and 30m in hydration TCPB tailings are both 102.5% of OCPB.

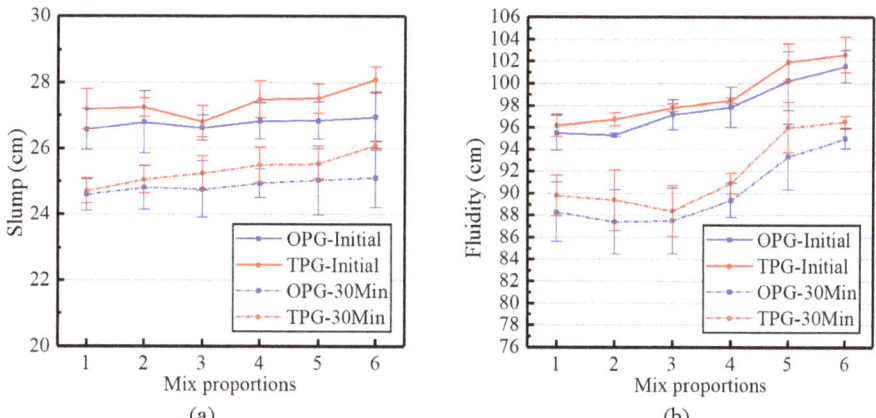

Figure 8. Workability of the mixtures with OPG and TPG. (**a**) Values of slump; (**b**) values of fluidity.

3.3. The UCS of PG-Based CPB

Compressive strength is one of the most important factors used to measure the quality of backfill. According to the previous publications, the required static strength of CPB curing for 28 days without exposures is arbitrarily selected at 0.2 MPa [51,52]. Besides, much of the research [24] on backfill tailings have determined that mixtures with PG have little strength in 7 days, but that the strength increases rapidly in the middle and later stages. In addition, studies [7,14] of receiver biases suggest that the required strength of PG-based CPB is about 1 MPa at 28 days curing time. Figure 9 summarises the evolution of the UCS values for the OCPB and TCPB samples which all increase with curing age, despite that the growth rate varies with different stages. The 7 days UCS of OCPB is only 0.140–0.555 MPa, however, the value of UCS increases significantly with increase of the mass concentration and cement content: eventually it reaches 0.499–1.111 MPa. Although the general rule for UCS growth is also applied to the evolution of TCPB, the UCS of TCPB is much greater than that of OCPB under the same mix proportion. The 7 days UCS of TCPB is from 0.482 MPa to 0.838 MPa. At 14 days, the strength values are from 0.710 MPa to 1.100 MPa and the values at 28 days climbed to the maximum range 0.825–2.531 MPa.

The ratios of 7, 14, and 28 days UCS values of OCPB/TCPB is shown in Figure 10. It can be seen that the ratio of 7 days UCS ranges from 28.9% to 67.3%, then the ratio gradually stabilises at nearly 65.5% with increase of the curing age. Furthermore, the ratio of mix-proportion four remains near 70%. One of the possible factors associated with this phenomenon is the cementitious activity of PG, which can replace OPC under certain conditions of alkaline activation. One of the possible factors [53] associated with the stronger retarding phenomena of the OCPB is the soluble phosphate in the OPG that reacted with Ca^{2+} ions and formed a thin film covering the cement clinker particles, which slowed down the early hydration process of the OPC. However, the film is broken with the hydration proceeds, and the later strength development will not be deteriorated. Therefore, according to the XRF results, soluble phosphorus in OPG separated from PG crystal with water-washing, as a result, this deterioration of mechanical properties of PG-based CPB was weakened. However, with the hydration reaction, the film in PG-based CPB is broken, and the 14 and 28 days UCS values of OCPB increase rapidly.

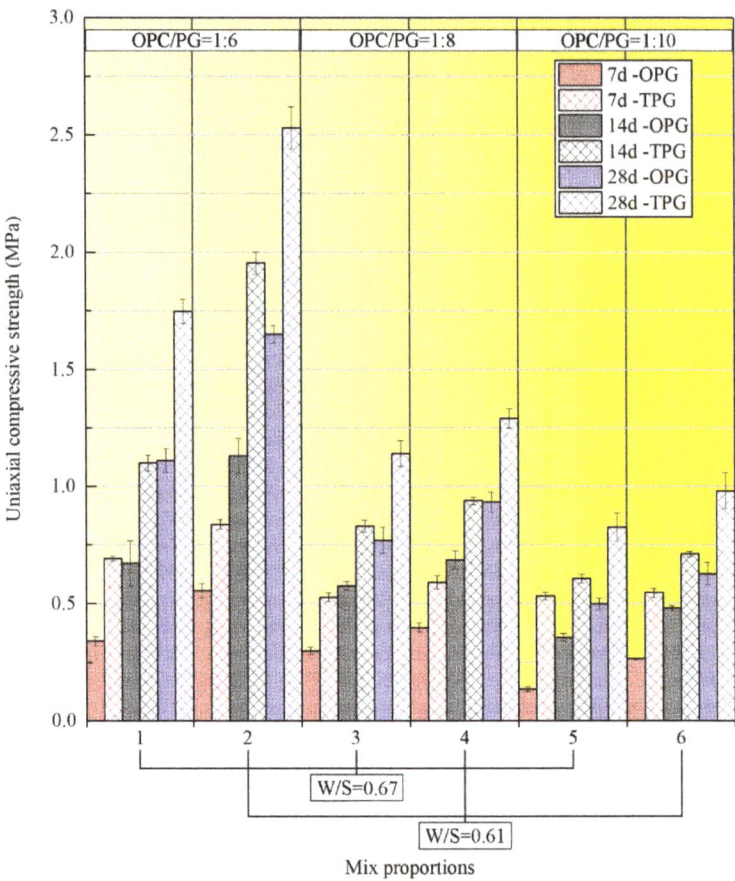

Figure 9. Uniaxial compressive strength (UCS) comparison of OPG-based CPB (OCPB) and treated PG (TPG)-based CPB (TCPB).

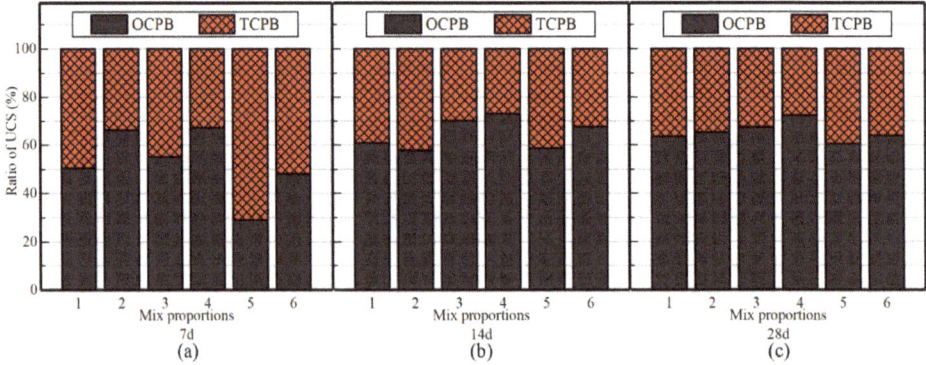

Figure 10. Growth rate ratio of OCPB and TCPB at (**a**) 7 days, (**b**) 14 days, and (**c**) 28 days.

In order to investigate the influence of OPG and TPG on strength growth of PG-based CPB in different curing ages, the strength growth rates of OCPB and TCPB at 0–7 days, 7—14 days and 14—28 days were calculated, using the 28 days UCS value of each sample as the final strength of this study, the ratios of UCS growth values in 0–7 days, 7–14 days and 14–28 days to final strength are shown in Figure 11. It can be summarised that the UCS values of OCPB increased slowly at the early ages of hydration, ranging from 27.96% to 42.55%, and rapidly in the later stage, which strongly confirms the analysis of the soluble phosphate's influence as noted above. By contrast, the values of TCPB all grew rapidly at the period of 0–7 days, ranging from 33.11% to 58.42%, and then gradually increased at the period of 7–28 days, which indicates that the strength formation of TCPB was mainly at 0–7 days. Therefore, water-washing pre-treatment can greatly reduce the retarding effect of PG on cement hydration in the early stage, however, with the hydration reaction, the mechanical property of PG-based CPB is still mainly determined by the mass concentration of mixture and the content of cement.

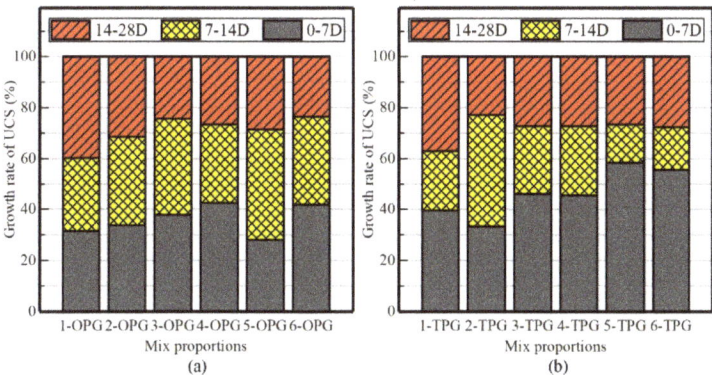

Figure 11. Growth rate ratio of UCS values of (**a**) OCPB, and (**b**) TCPB at different ages.

3.4. Microstructures of CPB

Figure 12 shows XRD patterns for the OCPB and TCPB (mix-proportion 3-OPG and 3-TPG) after 7, 14, and 28 days, the detected mineral phases are mainly $CaSO_4 \cdot 2H_2O$, AFt, and SiO_2. The probable reason that calcium silicate hydrate gel (C-S-H) cannot be detected in the XRD patterns is its amorphous or non-crystal micromorphology. As regards the $Ca(OH)_2$, this can react with the impurities of PG, making the diffraction peak difficult to distinguish (compared with the main diffraction peaks). The reaction Equations are as follows [54]:

$$P_2O_5 + 3Ca(OH)_2 \rightarrow Ca_3(PO_4)_2 \downarrow + 3H_2O \qquad (8)$$

$$2NaF + Ca(OH)_2 \rightarrow CaF_2 \downarrow + 2NaOH \qquad (9)$$

Based on the characteristic peaks of $CaSO_4 \cdot 2H_2O$, it can be concluded that the content of $CaSO_4 \cdot 2H_2O$ both decreases gradually with the hydration of these two hydration systems. What's more, compared with OCPB, the 7 days diffraction peaks of $CaSO_4 \cdot 2H_2O$ in TCPB are not that strong in the XRD patterns illustrating that substantial $CaSO_4 \cdot 2H_2O$ participates in the hydration system. However, as regards the 14, and 28 days diffraction peaks of $CaSO_4 \cdot 2H_2O$, TCPB are slightly stronger which may be associated with the inhibition effect of alkaline environment. A previous study [55] demonstrates that the hydration of $CaSO_4 \cdot 0.5H_2O$ and $CaSO_4 \cdot 2H_2O$ is inhibited by alkali, while a great quantity of acidic substances in TPG have been discharged with several water-washings that increased alkalinity of liquid phase in TCPB. Even so, the 7, 14, and 28 days diffraction peaks of AFt in TCPB were all greater than that of OCPB, which indicates that there were more hydration products and higher hydration degree in TCPB. Meanwhile, it can be observed that the 14, and 28 days characteristic peaks of Aft in OCPB increase significantly which further proves the disciplines of formation and destruction of the film, with the increase of curing time, the film in OCPB breaks and the hydration reaction continues. That is to say, water-washing pre-treatment can effectively enhance the composition of hydration products of PG-based CPB and stimulate the formation of AFt in early ages.

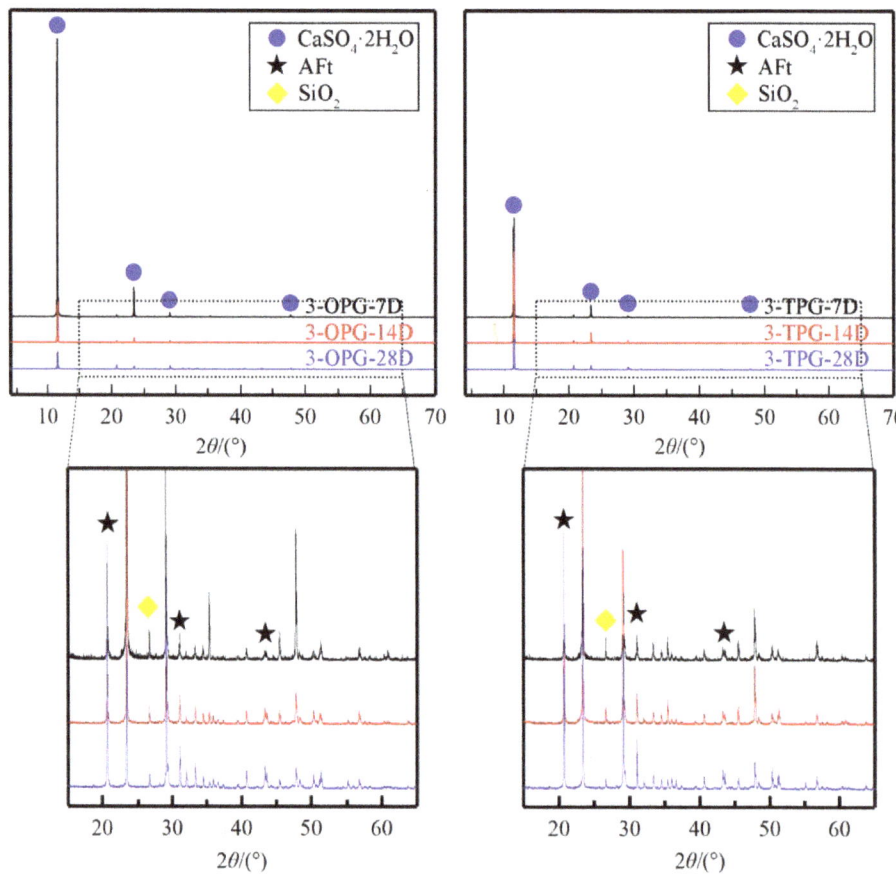

Figure 12. XRD patterns of OCPB and TCPB.

Subsequently, Figure 13 shows SEM images of the samples (Mix-proportion 3-OPG and 3-TPG) at different curing ages: left (Figure 13a,d) shows the morphology of specimens at 7 days, in the middle (Figure 13b,e) are specimens at 14 days, and on the right (Figure 13c,f) are at 28 days. These hydration products are needles, rods, or flocculants, among them, the needle or rod-like substances are ettringite (AFt). The floc material is CSH and covers the surfaces of the $CaSO_4 \cdot 2H_2O$ crystal particles. By comparing Figure 13a–c with Figure 13d–f, it can be seen that the microstructure of hydration products of TCPB was better developed than OCPB at any curing time. As shown in Figure 13a, after 7 days of hydration, the prismatic and slab-like PG crystals in OCPB remained visible, even within image of 14 days. Besides, the impurities absorbed on PG crystal surface nearly gone and small amount of relatively tiny AFt was produced in the pores between the hydration products and $CaSO_4 \cdot 2H_2O$ crystals. By contrast, from Figure 13d, although the pores of TCPB were not filled with hydration products, the microstructure was more compact with plenty of hydration products generating on the surface of PG particles. Figure 13c,f presents that a great deal of C-S-H gel both generated in OCPB and TCPB illustrating that the retarding effect of OPG decreased gradually with the hydration, while in OCPB, the generated C-S-H gel was much slenderer with huge pore size. Therefore, it can be concluded that water-washing pre-treatment can effectively improve the microstructure of PG-based CPB by promoting the early hydration reaction and optimising the porosity of hydration system which improves the mechanical properties.

Figure 13. SEM images of OCPB samples with (**a**) 7 days, (**b**) 14 days, and (**c**) 28 days curing age and TCPB samples with (**d**) 7 days, (**e**) 14 days, and (**f**) 28 days curing age.

4. Conclusions

This study investigated the feasibility of water-washing to pre-treat PG, aimed to improve the filling performance of PG-based CPB. Based on the considerable experimental results, the following conclusions were summarized as follows:

1. During the water-washing pretreatment, a certain number of soluble impurities on the surface of PG crystals could be removed by water-washing, and the morphology and size of the PG crystals were slightly optimized for CPB.
2. The water-washing pretreatment effectively improved the transport mobility and efficiency of the TCPB mixtures. The workability of the TCPB mixtures were greater than that of OCPB, which is beneficial to transport the TCPB to the stopes underground.
3. Moreover, because of the purification by water-washing, the impurities, which greatly hamper the hydration process, were separated from the OPG. The UCS value of TCPB was up to 2.45 times higher than that of OCPB samples. It can be concluded that the figures for TCPB increase significantly from stages of 0 to 7 days, and the purification capacity of water-washing by enhance the early UCS growth rate.
4. With the water-washing pretreatment, the diffraction peaks of AFt in the hydration system were also enhanced, although the diffraction peaks of some other hydration products were not found in the mineral phase due to the high PG content.

Author Contributions: Y.L., Q.Z. and Q.C. conceived the project; Y.L., Z.S. and Z.H. designed and performed the experiments; data curation, C.Q.; Y.L. wrote initial drafts of the work; Q.C. and Q.Z. wrote the final paper. All authors discussed the results and commented on the manuscript.

Funding: This essay is supported by the Fundamental Research Funds for the Central Universities of Central South University (2018zzts731) and Mittal Student Innovation Project of Central South University (201810533268).

Conflicts of Interest: The authors declare no conflict of interest.

References

1. Howard, S.F.a.B.J. IAEA Technical Reports Series No. 475, Guidelines for Remediation Strategies to Reduce the Radiological Consequences of Environmental Contamination IAEA Safety Reports Series No. 78, Radiation Protection and Management of NORM Residues in the Phosphate Industry. *J. Radiol. Prot.* **2013**, *33*, 491–495. [CrossRef]
2. Bisone, S.; Gautier, M.; Chatain, V.; Blanc, D. Spatial distribution and leaching behavior of pollutants from phosphogypsum stocked in a gypstack: Geochemical characterization and modeling. *J. Environ. Manag.* **2017**, *193*, 567–575. [CrossRef] [PubMed]
3. Suárez, S.; Roca, X.; Gasso, S. Product-specific life cycle assessment of recycled gypsum as a replacement for natural gypsum in ordinary Portland cement: Application to the Spanish context. *J. Clean. Prod.* **2016**, *117*, 150–159. [CrossRef]
4. Yang, J.; Liu, W.; Zhang, L.; Xiao, B. Preparation of load-bearing building materials from autoclaved phosphogypsum. *Constr. Build. Mater.* **2009**, *23*, 687–693. [CrossRef]
5. Yang, L.; Yan, Y.; Hu, Z. Utilization of phosphogypsum for the preparation of non-autoclaved aerated concrete. *Constr. Build. Mater.* **2013**, *44*, 600–606. [CrossRef]
6. Liang, H.; Zhang, P.; Jin, Z.; DePaoli, D. Rare Earth and Phosphorus Leaching from a Flotation Aggregate of Florida Phosphate Rock. *Minerals* **2018**, *8*, 416. [CrossRef]
7. Chen, Q.; Zhang, Q.; Fourie, A.; Xin, C. Utilization of phosphogypsum and phosphate aggregatefor cemented paste backfill. *J. Environ. Manag.* **2017**, *201*, 19–27. [CrossRef] [PubMed]
8. Abril, J.M.; Garcia-Tenorio, R.; Perianez, R.; Enamorado, S.M.; Andreu, L.; Delgado, A. Occupational dosimetric assessment (inhalation pathway) from the application of phosphogypsum in agriculture in South West Spain. *J. Environ. Radioactiv.* **2009**, *100*, 29–34. [CrossRef]
9. Yang, L.; Zhang, Y.; Yan, Y. Utilization of original phosphogypsum as raw material for the preparation of self-leveling mortar. *J. Clean. Prod.* **2016**, *127*, 204–213. [CrossRef]
10. Rajković, M.B.; Tošković, D.V. Phosphogypsum Surface Characterisation using Scanning Electron Microscopy. *Acta Periodica Technol.* **2003**, *34*, 61–70. [CrossRef]
11. Ali, M.A.; Lee, C.H.; Kim, S.Y.; Kim, P.J. Effect of industrial by-products containing electron acceptors on mitigating methane emission during rice cultivation. *Waste Manag.* **2009**, *29*, 2759–2764. [CrossRef]
12. Jiang, G.; Wu, A.; Wang, Y.; Lan, W. Low cost and high efficiency utilization of hemihydrate phosphogypsum: Used as binder to prepare filling material. *Constr. Build. Mater.* **2018**, *167*, 263–270. [CrossRef]
13. Chen, Q.; Zhang, Q.; Qi, C.; Fourie, A.; Xiao, C. Recycling phosphogypsum and construction demolition waste for cemented paste backfill and its environmental impact. *J. Clean. Prod.* **2018**, *186*, 418–429. [CrossRef]
14. Shen, W.; Gan, G.; Dong, R.; Chen, H.; Tan, Y.; Zhou, M. Utilization of solidified phosphogypsum as Portland cement retarder. *J. Mater. Cycles Waste* **2012**, *14*, 228–233. [CrossRef]
15. Yilmaz, E.; Belem, T.; Bussière, B.; Mbonimpa, M.; Benzaazoua, M. Curing time effect on consolidation behaviour of cemented paste backfill containing different cement types and contents. *Constr. Build. Mater.* **2015**, *75*, 99–111. [CrossRef]
16. Fall, M.; Benzaazoua, M.; Ouellet, S. Experimental characterization of the influence of aggregate fineness and density on the quality of cemented paste backfill. *Miner. Eng.* **2005**, *18*, 41–44. [CrossRef]
17. Cihangir, F.; Akyol, Y. Mechanical, hydrological and microstructural assessment of the durability of cemented paste backfill containing alkali-activated slag. *Int. J. Min. Reclam. Environ.* **2018**, *32*, 123–143. [CrossRef]
18. Ercikdi, B.; Cihangir, F.; Kesimal, A.; Deveci, H.; Alp, I. Utilization of water-reducing admixtures in cemented paste backfill of sulphide-rich mill Aggregate. *J. Hazard. Mater.* **2010**, *179*, 940–946. [CrossRef]
19. Cihangir, F.; Ercikdi, B.; Kesimal, A.; Ocak, S.; Akyol, Y. Effect of sodium-silicate activated slag at different silicate modulus on the strength and microstructural properties of full and coarse sulphidic aggregate paste backfill. *Constr. Build. Mater.* **2018**, *185*, 555–566. [CrossRef]
20. Dang, W.G.; Liu, Z.X.; He, X.Q.; Liu, Q.L.J.M.T. Mixture ratio of phosphogypsum in backfilling. *Min. Technol.* **2013**, *122*, 1–7. [CrossRef]
21. Shen, W.; Zhou, M.; Zhao, Q. Study on lime–fly ash–phosphogypsum binder. *Constr. Build. Mater.* **2007**, *21*, 1480–1485. [CrossRef]

22. Nizevičienė, D.; Vaičiukynienė, D.; Michalik, B.; Bonczyk, M.; Vaitkevičius, V.; Jusas, V.J.C.; Materials, B. The treatment of phosphogypsum with zeolite to use it in binding material. *Constr. Build. Mater.* **2018**, *180*, 134–142. [CrossRef]
23. Shen, Y.; Qian, J.; Huang, Y.; Yang, D. Synthesis of belite sulfoaluminate-ternesite cements with phosphogypsum. *Cem. Concr. Comp.* **2015**, *63*, 67–75. [CrossRef]
24. Li, X.; Du, J.; Gao, L.; He, S.; Gan, L.; Sun, C.; Shi, Y. Immobilization of phosphogypsum for cemented paste backfill and its environmental effect. *J. Clean. Prod.* **2017**, *156*, 137–146. [CrossRef]
25. Romero-Hermida, M.I.; Borrero-López, A.M.; Alejandre, F.J.; Flores-Alés, V.; Santos, A.; Franco, J.M.; Esquivias, L. Phosphogypsum waste lime as a promising substitute of commercial limes: A rheological approach. *Cem. Concr. Comp.* **2019**, *95*, 205–216. [CrossRef]
26. Taher, M.A. Influence of thermally treated phosphogypsum on the properties of Portland slag cement. *Resour. Conserv. Recycl.* **2007**, *52*, 28–38. [CrossRef]
27. Singh, M. Treating waste phosphogypsum for cement and plaster manufacture. *Cem. Concr. Res.* **2002**, *32*, 1033–1038. [CrossRef]
28. Mun, K.J.; Hyoung, W.K.; Lee, C.W.; So, S.Y.; Soh, Y.S. Basic properties of non-sintering cement using phosphogypsum and waste lime as activator. *Constr. Build. Mater.* **2007**, *21*, 1342–1350. [CrossRef]
29. Potgieter, J.H.; Potgieter, S.S.; McCrindle, R.I.; Strydom, C.A. An investigation into the effect of various chemical and physical treatments of a South African phosphogypsum to render it suitable as a set retarder for cement. *Cem. Concr. Res.* **2003**, *33*, 1223–1227. [CrossRef]
30. Kamimura, A.; Konno, E.; Yamamoto, S.; Watanabe, T.; Yamada, K.; Tomonaga, F. Improved method for the formation of recycled resins from depolymerized products of waste fiber-reinforced plastics: Simple and effective purification of recovered monomers by washing with water. *J. Mater. Cycles Waste* **2009**, *11*, 133–137. [CrossRef]
31. Slavinskaya, G.V. Water pretreatment to remove organic impurities and desalination with ion exchangers. *Russ. J. Appl. Chem.* **2003**, *76*, 1089–1093. [CrossRef]
32. Shuang, Y.; Hou, Y.; Zhang, B.; Yang, H.G. Impurity-Free Synthesis of Cube-Like Single-Crystal Anatase TiO2 for High Performance Dye-Sensitized Solar Cell. *Ind. Eng. Chem. Res.* **2013**, *52*, 4098–4102. [CrossRef]
33. Ali, M.A.; Kim, P.J.; Inubushi, K. Mitigating yield-scaled greenhouse gas emissions through combined application of soil amendments: A comparative study between temperate and subtropical rice paddy soils. *Sci. Total Environ.* **2015**, *529*, 140–148. [CrossRef]
34. Cárdenas-Escudero, C.; Morales-Flórez, V.; Pérez-López, R.; Santos, A.; Esquivias, L. Procedure to use phosphogypsum industrial waste for mineral CO_2 sequestration. *J. Hazard. Meter.* **2011**, *196*, 431–435. [CrossRef]
35. *General Purpose Portland Cement*; B175-2007; Standardization Administration of China: Beijing, China, 2007.
36. *Environmental Quality Standards for Surface Water*; GB 3838-2002; Standardization Administration of China: Beijing, China, 2002.
37. *Integrated Wastewater Discharge Standard*; GB 8978-1996; Standardization Administration of China: Beijing, China, 1996.
38. *Mould for Concrete Specimens*; JG237-2008; Ministry of Housing and Urban-Rural Construction of the People's Republic of China: Beijing, China, 2008.
39. Wu, A.; Wang, Y.; Wang, H.; Yin, S.; Miao, X. Coupled effects of cement type and water quality on the properties of cemented paste backfill. *Int. J. Min. Process.* **2015**, *143*, 65–71. [CrossRef]
40. Liu, L.; Fang, Z.; Qi, C.; Zhang, B.; Guo, L.; Song, K.I.I.L. Numerical study on the pipe flow characteristics of the cemented paste backfill slurry considering hydration effects. *Powder Technol.* **2019**, *343*, 454–464. [CrossRef]
41. Qi, C.; Chen, Q.; Fourie, A.; Zhao, J.; Zhang, Q. Pressure drop in pipe flow of cemented paste backfill: Experimental and modeling study. *Powder Technol.* **2018**, *333*, 9–18. [CrossRef]
42. *Standard Test Method for Slump of Hydraulic-Cement Concrete*; C143/C143M-2015; American Society for Testing and Materials International (ASTM): West Conshohocken, PA, USA, 2015.
43. Qi, C.; Chen, Q.; Fourie, A.; Tang, X.; Zhang, Q.; Dong, X.; Feng, Y. Constitutive modelling of cemented paste backfill: A data-mining approach. *Constr. Build. Mater.* **2019**, *197*, 262–270. [CrossRef]
44. *Standard Test Methods for Compressive Strength of Cylindrical Concrete Specimens*; C39 C39M-2014a; American Society for Testing and Materials International (ASTM): West Conshohocken, PA, USA, 2015.

45. Liu, T. Study on Pretreatment and Application of Phosphogypsum. Master's Thesis, Wuhan University, Wuhan, China, 2010.
46. Wang, A.Q.; Zhang, C.Z.; Zhang, N.S. The theoretic analysis of the influence of the particle size distribution of cement system on the property of cement. *Cem. Concr. Res.* **1999**, *29*, 1721–1726. [CrossRef]
47. Sari, D.; Pasamehmetoglu, A.G. The effects of gradation and admixture on the pumice lightweight aggregate concrete. *Cem. Concr. Res.* **2005**, *35*, 936–942. [CrossRef]
48. Yao, W. Research and Application of High Concentration Filling Theory of Coarse Aggregate in Mines. Ph.D. Thesis, Kunming University of Science and Technology, Kunming, China, 2011.
49. Yang, J.-C.; Wu, H.-D.; Teng, N.-C.; Ji, D.-Y.; Lee, S.-Y. Novel attempts for the synthesis of calcium sulfate hydrates in calcium chloride solutions under atmospheric conditions. *Ceram. Int.* **2012**, *38*, 381–387. [CrossRef]
50. Ma, B.; Lu, W.; Su, Y.; Li, Y.; Gao, C.; He, X. Synthesis of α-hemihydrate gypsum from cleaner phosphogypsum. *J. Clean. Prod.* **2018**, *195*, 396–405. [CrossRef]
51. Belem, T.; Benzaazoua, M. Design and application of underground mine paste backfill technology. *Geotech. Geol. Eng.* **2008**, *26*, 147–174. [CrossRef]
52. Li, M.; Moerman, A. Perspectives on the scientific and engineering principles underlying flow of mineral pastes. In Proceedings of the 34th Annual Meeting of CMP, Ottawa, ON, Canada, 22–24 January 2002; Paper No. 35. pp. 573–595.
53. Akın Altun, İ.; Sert, Y. Utilization of weathered phosphogypsum as set retarder in Portland cement. *Cem. Concr. Res.* **2004**, *34*, 677–680. [CrossRef]
54. Guo, C. Effect Mechanism of Fluorine and Phosphorus on Hydration Process of Cement Clinker. Ph.D. Thesis, Wuhan University of Technology, Wuhan, China, 2012.
55. Chen, X.; Gao, J.; Liu, C.; Zhao, Y. Effect of neutralization on the setting and hardening characters of hemihydrate phosphogypsum plaster. *Constr. Build. Mater.* **2018**, *190*, 53–64. [CrossRef]

 © 2019 by the authors. Licensee MDPI, Basel, Switzerland. This article is an open access article distributed under the terms and conditions of the Creative Commons Attribution (CC BY) license (http://creativecommons.org/licenses/by/4.0/).

Article

Durability Evaluation of Phosphogypsum-Based Cemented Backfill Through Drying-Wetting Cycles

Xibing Li [1], Shitong Zhou [1], Yanan Zhou [1], Chendi Min [1], Zhiwei Cao [1], Jing Du [1], Lin Luo [2] and Ying Shi [1,*]

[1] School of Resources and Safety Engineering, Central South University, Changsha 410083, Hunan, China; xbli@mail.csu.edu.cn (X.L.); zhoushitong@csu.edu.cn (S.Z.); yanan.zhou@csu.edu.cn (Y.Z.); mincdcsu@csu.edu.cn (C.M.); zhiweicao@csu.edu.cn (Z.C.); dujing040316@163.com (J.D.)

[2] Key Laboratory of Rock Mechanics and Geohazards of Zhejiang Province, Shaoxing 312000, Zhejiang, China; csulinluo1991@csu.edu.cn

* Correspondence: shiyingfriend@csu.edu.cn; Tel.: +86-186-7035-1208

Received: 28 April 2019; Accepted: 24 May 2019; Published: 26 May 2019

Abstract: In this study, the durability of phosphogypsum (PG)-based cemented backfill was investigated by drying-wetting cycles to explore deterioration of its strength and the release of impurities. The leachates in this test were composed of deionized water, 5% Na_2SO_4 solution, 5% NaCl solution, and a range of sulfuric acid solutions with pH values of 1.5, 3, and 5. After drying-wetting cycles, unconfined compressive strength (UCS), visual deterioration, porosity, microstructure and concentrations of phosphate and fluoride in the leachates were measured. The results showed that both saline and acidic solutions could lead to strength reduction of PG-based cemented backfill under different deterioration mechanisms. The mechanical damage of salinity was caused by micro-cracking and degradation of C–S–H. However, the H^+ broke the backfill by dissolving hydration products, leaving the conjunctures between PG particles weakened. Furthermore, the environmental impact was investigated by measuring the concentration of phosphate and fluoride in the leachates. In acidic solutions, the release of phosphate and fluoride was greatly enhanced by H^+. Compared to the great strength deterioration in saline leachates, the concentration of phosphate and fluoride were similar to that of deionized water, indicating that saline solutions had little impact on the release of hazardous impurities.

Keywords: drying-wetting cycles; durability; cemented phosphogypsum backfill; unconfined compressive strength; environment behavior

1. Introduction

A large number of cavities can be caused by the extraction of underground phosphate ore resources. Mining and industrial waste are often applied to fill the cavity of mines for the reduction of surface subsidence, increase of resource recovery and minimization of waste pollution [1–6]. Phosphogypsum (PG), as the main by-product of the phosphate fertilizer industry, mainly consists of $CaSO_4·2H_2O$ and some impurities. Previous studies have demonstrated the feasibility of recycling PG as a backfilling aggregate from both an engineering and environmental perspective [7,8]. In the PG-based backfill process, the slurry, consisting of PG, a hydraulic binder and water, has good pumping performance. When a PG-based backfill is refilled into cavities, the hardened structure of the backfill can provide early and cured strength [7,9].

Previous studies concentrate on the effect of raw materials (aggregate, binder, and water) on the cemented backfill process. However, after emplacement, the backfill is probably subject to unfavorable conditions during its long-term underground storage. Studies have shown that cementitious products can be damaged by high levels of salinity and acidity, leading to the expansion, cracking and even fracture of the hardened structure [10,11]. In the cement industry, factors affecting the evolution of mechanical properties have been extensively investigated. Firstly, sulfate attack is considered as an important factor influencing the mechanical performance of cemented materials due to the formation of gypsum and ettringite [11–15]. Expansive ettringite results in volume expansion and eventual cracks [16]. The expansion of secondary products can accelerate deterioration of the structure via penetration of harmful ions through the cracks [13,17]. Besides, chloride can affect the durability of the cement structure by converting hydration products into Friedel's salt [18,19]. Liu et al. concluded that an increase in chloride concentration has a detrimental influence on the unconfined compressive strength (UCS) of stabilized soil [19]. Furthermore, low pH can affect C–S–H by changing the C/S ratio [20]. Studies have shown that under the action of acid rain, backfill structures may be damaged or even completely collapse due to deterioration of cementitious materials [21]. However, the durability performance of backfill has not been clearly studied under various saline and acidic conditions. The mechanical properties of backfill are closely related to the safety of mining operations. Hence, the mechanical properties of backfill exposed to chemical erosion should be evaluated systematically.

As a typical solidification/stabilization (s/s) process, cemented backfill technology can effectively immobilize the impurities in PG, alleviating environmental pollution [22]. In the cement-based s/s process, chemical adsorption and physical adsorption are primary fixation mechanisms [23]. Chemical adsorption refers to the chemical precipitation of low solubility species, and usually occurs with high pH values [23]. Besides, impurities are physically encapsulated by hydration products, such as C–S–H. Although impurities can be stabilized by cementitious products, it is worth noting that cement-based solidified/stabilized impurities are vulnerable to external physical and chemical degradation processes [24,25]. A previous study has pointed out that high NaCl content can increase the leachability of hazardous components in stabilized soil [19]. Similarly, leaching of toxic impurities from cement-stabilized soil increases noticeably in an acidic environment [25]. However, studies on the leaching of hazardous impurities from PG-based backfill are still limited [7,22]. Therefore, research on the dynamics of impurities in backfill should be conducted under unfavorable conditions.

This study aimed to experimentally analyze the durability and environmental risks of PG-based backfill. Hydro-mechanical fatigue was used to accelerate the aging of cemented PG-based backfill. As a common method for durability analysis, a drying-wetting test was conducted in this study. Changes in the mechanical strength and pollutant contents in leachates were the main research themes. To this end, PG-based backfill specimens with a curing age of 120 d were exposed to different solutions (deionized water, 5% Na_2SO_4 solution, 5% NaCl solution, and a range of acidic solutions with pH values of 1.5, 3, and 5). Samples were then taken after 1, 3, 5, 10, 15, 21 and 30 drying-wetting cycles. Then, the physical, mechanical and microstructural properties were evaluated, including the UCS, morphology, mass, porosity, and microstructure. Simultaneously, the pH fluctuation and the concentration of phosphate and fluoride in the leachates were measured and analyzed.

2. Materials and methods

2.1. Raw Material and the Preparation of Backfill Specimens

The experimental PG and binder were purchased from Guizhou Kailin (Group) Co., Ltd., Guiyang, China. The composite binder was produced in mines by mixing and grinding yellow phosphorous slag, calcareous material and cement clinker at a ratio of 6:1:3 [26]. The particle size distributions of the PG and binder were determined by using a particle size analyzer (Mastersizer 2000, Malvern, Malvern, UK), the results of which are shown in Figure 1. The coefficient of uniformity (C_u) and coefficient of curvature (C_c) were used to assess the distribution of particle size in the PG, as shown in Table 1. When PG is used as an aggregate of backfill, the smaller particle size of PG may result in lower strength than typical mine backfill [8]. The specific gravity values of the PG and binder used in this study were 2.35 and 3.21, respectively. The chemical compositions of the PG and binder were measured by X-ray fluorescence (S4 Pioneer, Bruker, Germany), with the results shown in Table 1.

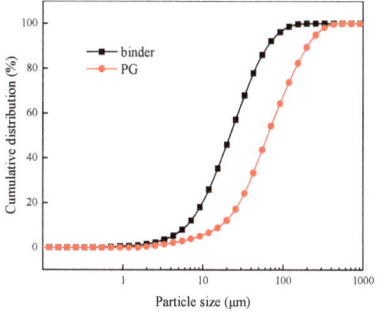

Figure 1. Particle size distributions of the phosphogypsum and binder.

Table 1. Chemical compositions and physical characteristics of the phosphogypsum and binder.

Chemical Composition	PG	Binder
	%	%
SiO_2	1.7	23.4
Fe_2O_3	0.3	2.6
CaO	35.9	54.4
MgO	0.1	1.7
SO_3	50.8	5.5
P_2O_5	2.6	1.7
Na_2O	0.1	1.3
K_2O	0.1	0.9
TiO_2	0.1	0.5
Physical Characteristic		
D_{10} (μm)	17.51	6.31
D_{30} (μm)	42.79	13.56
D_{60} (μm)	92.05	29.18
$C_u = D_{60}/D_{10}$	5.26	
$C_c = D_{30}^2/(D_{60} \times D_{10})$	1.14	
Specific gravity	2.35	3.21

PG, binder and deionized water (at a ratio of 4:1:5 by weight fraction) were mixed at a rotating speed of 200 rpm for 30 min. The backfill slurry had a solid concentration of 50% and a binder dosage of 20%. Compared to the 3~7% of binder used for common cemented paste backfill [7], a high binder percentage was adopted in this study (20%), which was rationalized as follows: Firstly, the high content of acid and sulfate in the PG could have led to a higher demand for binder. Secondly, a reduction in the strength of the cemented materials could have resulted from a lower specific area and lower hydraulic activity of the binder, and poor design, etc. Therefore, a high binder content was adopted in this study to ensure the performance of the PG-based backfill. The apparent viscosity of the slurry was 460 mPa·s, measured by a digital viscometer (NDJ-9S, Fangrui, China). Then, the homogeneous backfill slurry was cast into a plastic mold of dimensions 40 mm × 40 mm × 40 mm. After the initial setting, specimens were coded and placed in a chamber at a temperature of 20 ± 1 °C and a humidity of 95 ± 1%.

2.2. Drying-Wetting Test

A drying-wetting test was adopted to accelerate the deterioration of the PG-based backfill specimens. To reduce the hydration effect on degradation, the drying-wetting test was conducted after 120 d of curing, as shown in Figure 2. Extremely high salinity and acidity were chosen to accelerate the degradation process. Therefore, deionized water, 5% Na_2SO_4 solution, 5% NaCl solution, and three dilute sulfuric acid solutions with pH values of 1.5, 3, and 5 were used as exposure solutions in this study.

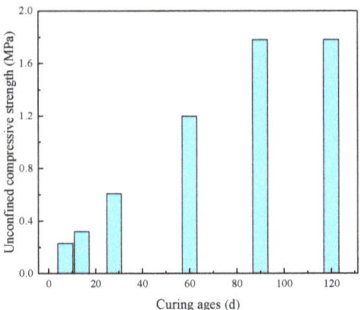

Figure 2. The development of unconfined compressive strength of the phosphogypsum (PG)-based backfill specimens within 120 days.

The drying-wetting test performed in this study was in accordance with the recommended method [11,27,28]. Firstly, the PG-based backfill specimens were dried in an oven at 40 °C for 48 h. Secondly, the specimens were submerged in different exposure solutions with a liquid/solid ratio of 5:1 for 24 h at an environment temperature of 20 °C. The specimens were then exposed to 1, 3, 5, 10, 15, 21 and 30 drying-wetting cycles, with the leachate sampled after the specified number of cycles and filtered (0.22 μm) for subsequent chemical analysis.

2.3. UCS Tests

A UCS test is a convenient method for monitoring the quality of backfill [29]. As such, the UCS of the specimens with 1, 3, 6, 10, 15, 21 and 30 drying-wetting cycles was measured. A rigid hydraulic pressure servo machine of 10 kN loading capability (WHY200/10, Hualong, Shanghai, China) was used to test the UCS of the specimens at a constant displacement rate of 0.1 mm/min. To avoid randomness and contingency in the test data, each test was performed three times and the mean values obtained for subsequent analysis. According to the mine operators of Guizhou Kailin (Group) Co., Ltd., the strength of PG-based backfill should be more than 1.0 MPa to satisfy the needs of safe mining operations.

2.4. Chemical Analysis

The pH values of leachates were measured using a pH meter (STARTER300, Ohaus, Parsippany, NJ, America). The concentration of F^- was measured by a fluorine ion-selective electrode (PF-1-01, Leici, Shanghai, China). The concentrations of SO_4^{2-} and dissolved PO_4^{3-} were tested via a spectrophotometer (UV1800, Shimadzu, Kyoto, Japan). Specifically, barium chromate spectrophotometry was employed to evaluate SO_4^{2-} concentration in the leachates, while the concentration of dissolved PO_4^{3-} was determined by ammonium molybdate tetrahydrate spectrophotometry.

2.5. Porosity Measurement

Volume accuracy of the specimen is usually required in typical methods of porosity measurement [30]. However, due to spalling on the surface of specimens, volume testing may be inaccurate and result in measurement errors of the total porosity. Therefore, the porosity change was evaluated as follows:

$$p = \frac{m_{n-s} - m_{n-d}}{m_{n-s}} \times 100\%, \tag{1}$$

where p was the porosity of the specimen; m_{n-s} was the saturated mass of the specimen after n cycles; and m_{n-d} was the dry mass of the specimens after n cycles. Mass changes in the specimens were measured before and after treatment with the exposure solutions by electronic balance with a capacity of 2 kg and an accuracy of 0.01 g.

2.6. Microstructural Studies

Scanning electron microscopy (SEM) and energy dispersive spectrometry (EDS) were conducted to evaluate the microstructural development of the PG-based cemented backfill using the HELIOS NamoLab 600i (FEI, Lake Oswego, America). To prevent further hydration, fractured pieces of PG-based backfill specimens were immediately soaked in an ethanol solution after UCS tests [7]. Prior to the SEM analysis, fractured pieces were dried in an oven at 40 °C. Owing to their poor conductivity, broken specimens were covered with a layer of gold coating in a vacuum to meet the inspection requirements. SEM and EDS analyses were operated at a magnification factor of 2500, an accelerating voltage of 10.00 kV, and a working distance of 6.0 mm.

3. Results and Discussion

3.1. Properties of Backfill Exposed to Chemical Solutions

3.1.1. Visual Assessment

Figure 3 shows the appearance of the PG-based backfill specimens after 30 drying-wetting cycles. Specimens exposed to deionized water were visually intact, while deterioration was observed for specimens under other conditions (in both acidic and saline conditions). When the specimens were soaked in acidic solutions, there was obvious spalling and pores on the surface of the specimens. When specimens were exposed to saline solutions, there were quite different destruction modes, with salting-out and microcracks observed. Due to the recrystallization of saline, a great number of white crystals appeared on the surface of the backfill specimens. To some extent, the crystallization pressure contributes to the formation of microcracks [31]. Notably, when the specimens were exposed to 5% Na_2SO_4, there was an obvious propagation of microcracks, resulting from the expansion of secondary gypsum and secondary ettringite. Presumably, the large amount of SO_4^{2-} in the solution combines with Ca^{2+} derived from C–H or C–S–H to form the secondary gypsum [32–35]. At the same time, SO_4^{2-} may also promote the continued growth of ettringite [36]. The expansive properties of these secondary products seemingly enhance the formation of microcracks in the backfill.

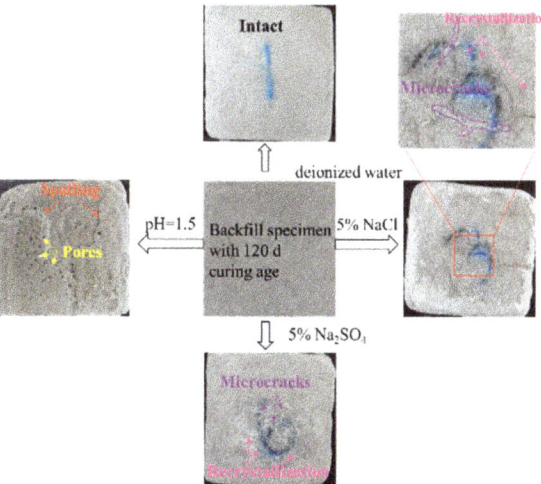

Figure 3. Visual inspection of the PG-based backfill specimens with different exposure solutions after 30 drying-wetting cycles.

3.1.2. Microstructure

The microstructure of the deteriorated specimens was studied by SEM–EDS, as shown in Figure 4. For the specimens cured for 120 days under normal conditions (at the temperature of 20 °C and humidity of 95%), the hydration process proceeded smoothly, and the PG particles were surrounded by a large amount of hydration products, including ettringite and C–S–H. However, for the specimens subjected to acidic and saline solutions, there were some morphological changes after 30 drying-wetting cycles. In the acidic environment (pH = 1.5), a large number of naked PG particles were observed, which may be related to the dissolution of hydration products [37]. As a result, conjunctions of the PG particles were broken, resulting in spalling of the PG-based backfill from a macroscopic perspective. In the NaCl solution, many regular crystals among the PG particles were observed, with EDS analysis indicating that the regular crystals were recrystallized NaCl (as presented in Figure 5). A large amount of recrystallized salt was observed in the SEM images, which was consistent with the visual observation results. When the specimens were submerged in 5% NaCl solution, salt recrystallization inside the backfill generated disruptive pressure and contributed to the propagation and emergence of microcracks, as shown in Figure 3. For the specimens submerged in Na_2SO_4 solution, a great amount of needle-like ettringite and recrystallized Na_2SO_4 was observed between the PG particles. As the volume of secondary ettringite and recrystallized Na_2SO_4 increased [38], it induced crystallization pressure, culminating in the cracking of the backfill specimens.

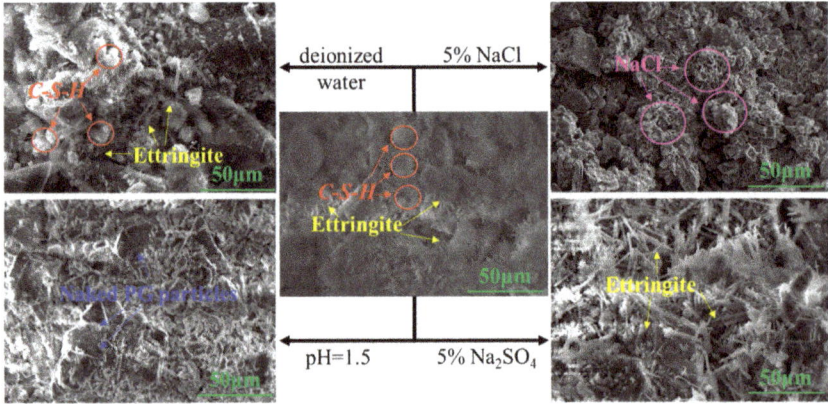

Figure 4. Scanning electron microscope (SEM) images of the PG-based backfill specimens with 120 d curing age and different exposure solutions after 30 drying-wetting cycles. Magnification factor: 2500×; accelerating voltage: 10.00 kV; and working distance: 6.0 mm.

Figure 5. Energy dispersive spectrometry (EDS) analysis of specimens exposed to NaCl. Magnification factor: 2500×; accelerating voltage: 10.00 kV; and working distance: 6.0 mm.

3.1.3. Pore Structure

Pore structure has a significant influence on the mechanical properties of cemented backfill [39,40]. Thus, the porosity of the backfill specimens was measured to evaluate the mechanical performance of the PG-based backfill. In this study, the porosity was characterized by the mass ratio of pore water and the saturated specimen, as shown in Equation (1).

Figure 6 shows the porosity of the backfill specimens after 30 drying-wetting cycles. The porosity of specimens exposed to deionized water after the first drying-wetting cycle (23.55%) could be interpreted as the initial state, as shown by the dotted line in Figure 6. It may be observed that in deionized water, the porosity of the specimens increased slightly with the extension of drying-wetting cycles. A probable reason for this is the original hydration products that filled in the pores were gradually dissolved.

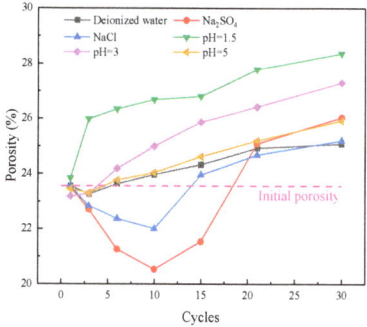

Figure 6. Variation of the porosity of hardened backfill samples with different drying-wetting cycles.

On the other hand, the porosity of the specimens increased significantly in acidic conditions. After 30 total drying-wetting cycles, specimens exposed to acidic conditions showed a gradual increase in porosity with an increase in the acidity of the leachates, which was highly noticeable in comparison to the specimen submerged in deionized water. This increase in porosity in acidic conditions may be explained as follows: Firstly, the presence of more H^+ was able cause greater dissolution of the hydration products, resulting in the exposure of naked PG—as presented in Figure 4. Secondly, the solubility of the aggregate ($CaSO_4 \cdot 2H_2O$) increased with the increase in H^+, resulting in more voids in the backfill. Hence, there was a higher porosity in the acidic condition.

When the specimens were immersed in saline solutions, the porosity of the specimens first decreased and then increased. In the first ten cycles, the continuous recrystallization of the saline (Na_2SO_4 and NaCl) filled up tiny pores in specimens, probably generating the observed decrease in porosity. With the formation of secondary gypsum and ettringite, the porosity of specimens decreased more significantly in the Na_2SO_4 solution [41]. After ten cycles, however, the porosity increased rapidly with the extension of drying-wetting tests, as shown in Figure 6. This increase in porosity (presented in Figure 6) and the appearance of microcracks (presented in Figure 3) were probably caused by expansive cracks of the recrystallized saline and secondary products.

3.1.4. Strength Evolution

To analyze the strength evolution of the backfill, UCS tests were conducted [42,43]. Figure 7 compares the UCS results of the PG-based backfill specimens exposed to different leaching solutions after 1, 3, 6, 10, 15, 21 and 30 drying-wetting cycles. In general, the results of the visual assessment were in good agreement with the UCS tests. As shown in Figure 8, the more severe the visual damage on the surface of the PG-based backfill specimens, the lower the strength of specimens. Specimens in deionized water were visually intact with little change in UCS evolution, whereas both in saline and acid solution, PG-based backfill specimens exhibited visual damage and a decrease in their mechanical strength.

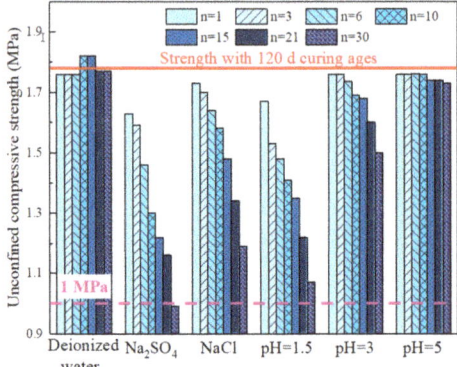

Figure 7. Unconfined compressive strength of the PG-based backfill specimens at different drying-wetting cycles and exposure solutions.

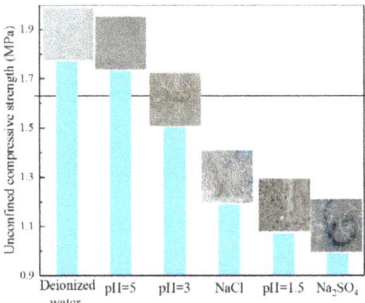

Figure 8. Visual inspection of the PG-based backfill specimens with different exposure solutions after 30 drying-wetting cycles.

When exposed to the saline solutions, the UCS of the specimens decreased significantly by about 40%. The strength reduction in saline solutions may be explained as follows: Firstly, the large amount of Na^+ in the leachates potentially replaces the Ca^{2+} in C–S–H, which lowered the Ca/Si ratio and resulted in strength reduction of C–S–H [44–46]. Secondly, as mentioned above, disruptive pressures generated by recrystallized saline (NaCl and Na_2SO_4) in pores resulted in the development of microcracks. As a result, the structure was broken with the lowered strength of the specimens [47].

Noticeably, UCS reduction of specimens exposed to Na_2SO_4 was more pronounced than those exposed to NaCl. This was likely caused by the large formation of expansive ettringite in Na_2SO_4 solution. As shown in Figure 4, a great deal of ettringite induced by SO_4^{2-} can enhance the crystallization pressure, leading to the formation of large microcracks. This result suggests that dissolved SO_4^{2-} is an important factor in accelerating the strength deterioration of backfill. This result was also consistent with a previous study, which reported that a 48% drop in UCS is observed when concrete is exposed to Na_2SO_4 solution for 22 months [48].

When the specimens were exposed to acidic solutions, the presence of more H^+ appeared to lead to a greater loss of UCS. Chen et al. concluded that acid rain attack can decrease the UCS of cementitious materials by 34.2% [21]. In this study, the UCS of the backfill decreased by 39.9%, 15.7%, and 2.81% for the acidic exposure solutions of pH 1.5, 3 and 5, respectively (after 30 drying-wetting cycles). This decrease in strength was probably caused by the dissolution of hydration products, which plays a key role in strength enhancement. Firstly, as shown in Figure 4, naked PG particles indicated the dissolution of hydration products. Secondly, the low pH value of pore water was able to directly inhibit the hydration processes, resulting in decreased strength [49]. In addition to the decrease of hydration products, the dissolution of the aggregate ($CaSO_4 \cdot 2H_2O$) contributed to strength reduction. The solubility of $CaSO_4 \cdot 2H_2O$ increased with the decrease in pH values, resulting in structural deterioration and the decrease in the strength of the backfill.

Studies have shown that, in a mine, adequate UCS is about 0.7–2.0 MPa [50,51], and the required strength varies largely in line with the function and application of the backfill. According to mine operators, adequate strength of PG-based backfill is about 1.0 MPa. As shown in Figure 8, the lowest residual strength is 0.99 MPa (Na_2SO_4, 30 cycles). In this case, the PG-based backfill can satisfy the safety requirements for operation in mines, even when subjected to extremely unfavorable conditions (e.g., high salinity or acidity).

3.2. Impurity Dynamics Through Drying-Wetting Cycles

3.2.1. Fluctuation of pH Values

It is well known that pH has an important influence on the hydration process of cemented backfill [49]. Therefore, the pH values of leachates during 30 drying-wetting cycles were measured to assess the hydration process of the PG-based backfill, the results of which are presented in Figure 9. The pH value of each leachate was higher than that of its original solution. When the specimens were immersed in deionized water, alkaline cementitious agents led to the alkalescence of the leachates. In the acidic environment, hydration products, such as C–H and C–S–H, may have reacted with H^+, resulting in a pH higher than that of the original solutions [21]. For the specimens exposed to saline solution, the pH value of the leachates was higher than that of the specimens exposed to deionized water, since SO_4^{2-} and Cl^- could react as follows [19,52]:

$$Ca(OH)_2 + 2SO_4^{2-} + 2H_2O = CaSO_4 \cdot 2H_2O + 2OH^- \text{a} = 1, \quad (2)$$

$$Ca(OH)_2 + 2Cl^- = CaCl_2 + 2OH^- \quad (3)$$

resulting in the release of OH^- and a rapid increase in pH values. Generally speaking, the pH value of each leachate decreased as the drying-wetting cycles proceeded, presumably due to the consumption of alkaline substances in the backfill specimens [7].

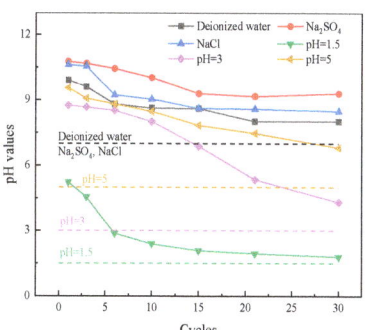

Figure 9. pH values in 30 drying-wetting cycles; the dotted lines indicate the initial values for each solution.

3.2.2. Dynamics of Phosphate and Fluoride

Previous studies have focused on the mechanical performance of mine backfill. However, PG, the aggregate of PG-based backfill, contains some impurities, such as phosphate and fluoride, which may cause serious environmental pollution [53]. Previous studies have shown that impurities originating from PG can be effectively fixed during the backfill process [7,22]. In PG-based backfill slurry, fluoride reacts with calcium to form precipitated Ca–F compounds (such as CaF_2) and Ca–P compounds (such as $Ca_3(PO_4)_2$ and $CaHPO_4$ [8]. However, these solidified impurities may be gradually released over time under unfavorable conditions [54]. Thus, a crucial point of reusing PG as an aggregate is to assess the chemical stability of the impurities in the PG-based backfill, since the backfill would be affected by underground conditions for a long period.

The pollution risk of phosphate and fluoride to water bodies has been emphasized recently. Figure 10 compares the cumulative leaching quantities of fluoride and dissolved phosphate in different the leachates. Generally, these findings suggest that solidified components can be released under unfavorable conditions after drying-wetting cycles, while the release quantities of phosphate and fluoride is related to the solution type.

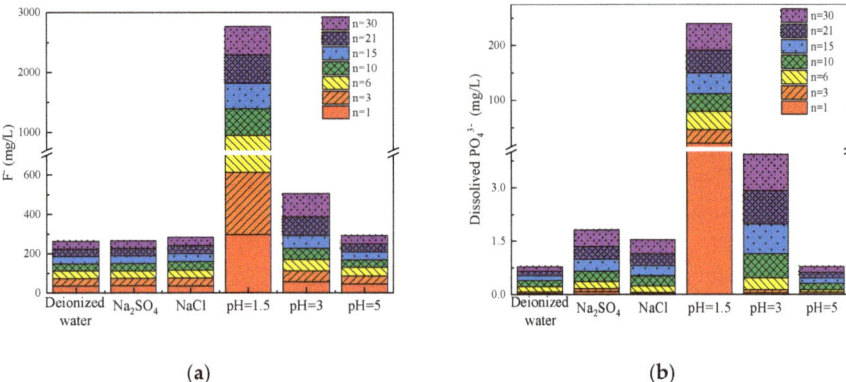

Figure 10. The cumulative leaching quantities of hazardous impurities in leachates: (**a**) The cumulative leaching quantities of fluoride and (**b**) the cumulative leaching quantities of dissolved phosphate.

As shown in Figure 10, a low pH has the largest impact on the release of fluoride and phosphate. The cumulative quantity of fluoride is increased by over 10 times when specimens are exposed to an acidic environment. Undissolved CaF_2 is dissolved slightly by H^+, leading to the release of fluoride. According to Figure 10b, in terms of phosphate, the cumulative quantity of dissolved phosphate leached by an acidic solution (pH = 1.5) is almost 150 times higher than that leached by deionized water. The sharp increase observed in the content of dissolved phosphate may be explained by the following reasons: Firstly, the dissolution of precipitated Ca–P compounds is an important factor. Secondly, the lattice constant of eutectic phosphate ($CaHPO_4 \cdot 2H_2O$) is similar to that of $CaSO_4 \cdot 2H_2O$, indicating that $CaHPO_4 \cdot 2H_2O$ enters the $CaSO_4 \cdot 2H_2O$ lattice to form a eutectic phosphate. As the $CaSO_4 \cdot 2H_2O$ lattice is seemingly destroyed by H^+, eutectic phosphate is released into the surrounding environment, leading to an increase in the content of dissolved phosphate.

Saline solution had little influence on the release of fluoride, as shown in Figure 10a. The cumulative fluoride quantities in both Na_2SO_4 and NaCl solutions were similar to those in deionized water. However, the release of phosphate in saline solutions was slightly higher than that of deionized water (almost 1.8 mg/L in saline solutions and 0.8 mg/L in deionized water) due to the damage of $CaSO_4 \cdot 2H_2O$ crystals. In addition, a portion of phosphate in the form of eutectic phosphate appeared to dissolve out of the gypsum lattice dissolution.

4. Conclusions

Through a drying-wetting test of 30 cycles, this study examined the mechanical change of PG-based backfill and the environmental dynamics of the pollutants. Based on the test results, a number of conclusions were drawn.

Firstly, in terms of the environment, high acidity had the largest impact on the release of impurities, since solidified components were dissolved by H^+. In comparison, the presence of environmental saline solution had a slight effect on the release of impurities. Secondly, from the perspective of strength evolution, both salinity and acidity led to significant strength reduction in the backfill under different mechanisms. In the acidic environment, the dissolution of hydration products was the main reason for the reduction in strength. In the saline solution, the strength of the backfill specimens was lowered due to the expansion of micro-cracks and the weakening of hydration products. Although significant strength reduction was observed under unfavorable conditions, PG-based backfill can maintain adequate strength for the safe operation of mines. Therefore, the utilization of PG-based backfill technology is an effective method to fill the cavity of mines.

Author Contributions: X.L. and Y.S. conceived and designed the theoretical framework; S.Z. performed the experiments and wrote the manuscript; Y.Z.; C.M.; Z.C.; J.D.; and L.L. corrected the tables and figures. All authors participated in the finalization of the written manuscript. X.L. and Y.S. acted as the supervisors of the project and acquired all the necessary funding.

Funding: This work was supported by the Project of Key Research Development Program of Hunan (Grant No. 2017SK2251), Hunan Natural Science Foundation (Grant No. 2018JJ3664) and the Fundamental Research Funds for the Central Universities of Central South University, China (Grant No. 2018zzts724 and 2019zzts305).

Acknowledgments: We thank Modern Analysis and Testing Center of CSU for assistance with SEM analysis.

Conflicts of Interest: The authors declare no conflict of interest.

References

1. Wu, D.; Hou, Y.B.; Deng, T.F.; Chen, Y.Z.; Zhao, X.L. Thermal, hydraulic and mechanical performances of cemented coal gangue-fly ash backfill. *Int. J. Miner. Process.* **2017**, *162*, 12–18. [CrossRef]
2. Jiang, H.Q.; Fall, M.; Liang, C. Yield stress of cemented paste backfill in sub-zero environments: experimental results. *Miner. Eng.* **2016**, *92*, 141–150.
3. Kesimal, A.; Yilmaz, E.; Ercikdi, B.; Alp, I.; Deveci, H. Effect of properties of tailings and binder on the short-and long-term strength and stability of cemented paste backfill. *Mater. Lett.* **2005**, *59*, 3703–3709. [CrossRef]
4. Wang, S.F.; Li, X.B.; Wang, S.Y. Three-dimensional mineral grade distribution modelling and longwall mining of an underground bauxite seam. *Int. J. Rock. Mech. Min.* **2018**, *103*, 123–136. [CrossRef]
5. Guo, G.L.; Zhu, X.J.; Zha, J.F.; Wang, Q. Subsidence prediction method based on equivalent mining height theory for solid backfilling mining. *T. Nonferr. Metal. Soc.* **2014**, *24*, 3302–3308. [CrossRef]
6. Ma, D.; Duan, H.Y.; Liu, J.F.; Li, X.B.; Zhou, Z.L. The role of gangue on the mitigation of mining-induced hazards and environmental pollution: An experimental investigation. *Sci. Total Environ.* **2019**, *664*, 436–448. [CrossRef] [PubMed]
7. Li, X.B.; Du, J.; Gao, L.; He, S.Y.; Gan, L.; Sun, C.; Shi, Y. Immobilization of phosphogypsum for cemented paste backfill and its environmental effect. *J. Clean. Prod.* **2017**, *156*, 137–146. [CrossRef]
8. Chen, Q.S.; Zhang, Q.L.; Qi, C.C.; Fourie, A.; Xiao, C.C. Recycling phosphogypsum and construction demolition waste for cemented paste backfill and its environmental impact. *J. Clean. Prod.* **2018**, *186*, 418–429. [CrossRef]
9. Ercikdi, B.; Cihangir, F.; Kesimal, A.; Deveci, H.; İbrahim, A. Utilization of water-reducing admixtures in cemented paste backfill of sulphide-rich mill tailings. *J. Hazard Mater.* **2010**, *179*, 940–946. [CrossRef]
10. Jiang, H.Q.; Fall, M. Yield stress and strength of saline cemented tailings in sub-zero environments: portland cement paste backfill. *Int. J. Miner. Process.* **2017**, *160*, 68–75. [CrossRef]
11. Helson, O.; Eslami, J.; Beaucour, A.L.; Noumowe, A.; Gotteland, P. Durability of soil mix material subjected to wetting/drying cycles and external sulfate attacks. *Constr. Build. Mater.* **2018**, *192*, 416–428. [CrossRef]
12. Wojciech, P. Analysis of carbonate and sulphate attack on concrete structures. *Eng. Fail. Anal.* **2017**, *79*, 606–614.
13. Kunther, W.; Lothenbach, B.; Scrivener, K.L. On the relevance of volume increase for the length changes of mortar bars in sulfate solutions. *Cement Concrete Res.* **2013**, *46*, 23–29. [CrossRef]
14. Fall, M.; Pokharel, M. Coupled effects of sulphate and temperature on the strength development of cemented tailings backfills: Portland cement-paste backfill. *Cement Concrete Res.* **2010**, *32*, 819–828. [CrossRef]

15. Cihangir, F.; Ercikdi, B.; Kesimal, A.; Ocak, S.; Akyol, Y. Effect of sodium-silicate activated slag at different silicate modulus on the strength and microstructural properties of full and coarse sulphidic tailings paste backfill. *Constr. Build. Mater.* **2018**, *185*, 555–566. [CrossRef]
16. Zhang, J.R.; Sun, M.; Hou, D.S.; Li, Z.J. External sulfate attack to reinforced concrete under drying-wetting cycles and loading condition: Numerical simulation and experimental validation by ultrasonic array method. *Constr. Build. Mater.* **2017**, *139*, 365–373. [CrossRef]
17. Cihangir, F.; Akyol, Y. Mechanical, hydrological and microstructural assessment of the durability of cemented paste backfill containing alkali-activated slag. *Int. J. Min. Reclam. Env.* **2016**, *185*, 1–21. [CrossRef]
18. Wang, Q.; Yan, P.Y.; Yang, J.W.; Zhang, B. Influence of steel slag on mechanical properties and durability of concrete. *Constr. Build. Mater.* **2013**, *47*, 1414–1420. [CrossRef]
19. Liu, J.J.; Zha, F.S.; Xu, L.; Yang, C.B.; Chu, C.F.; Tan, X.H. Effect of chloride attack on strength and leaching properties of solidified/stabilized heavy metal contaminated soils. *Eng. Geol.* **2018**, *246*, 28–35. [CrossRef]
20. Ragoug, R.; Metalssi, O.O.; Barberon, F.; Torrenti, J.M.; Roussel, N.; Divet, L.; Jean-Baptiste, E.L. Durability of cement pastes exposed to external sulfate attack and leaching: Physical and chemical aspects. *Cement Concrete Res.* **2019**, *116*, 134–145. [CrossRef]
21. Chen, M.C.; Wang, K.; Xie, L. Deterioration mechanism of cementitious materials under acid rain attack. Title of the article. *Eng. Fail. Anal.* **2013**, *27*, 272–285. [CrossRef]
22. Shi, Y.; Gan, L.; Li, X.B.; He, S.Y.; Sun, C.; Gao, L. Dynamics of metals in backfill of a phosphate mine of guiyang, China using a three-step sequential extraction technique. *Chemosphere* **2018**, *192*, 354–361. [CrossRef]
23. Chen, Q.Y.; Tyrer, M.; Hills, C.D.; Yang, X.M.; Carey, P. Immobilisation of heavy metal in cement-based solidification/stabilisation: a review. *Waste Manag.* **2009**, *29*, 390–403. [CrossRef] [PubMed]
24. Du, Y.J.; Wei, M.L.; Reddy, K.R.; Wu, H.L. Effect of carbonation on leachability, strength and microstructural characteristics of KMP binder stabilized Zn and Pb contaminated soils. *Chemosphere* **2015**, *144*, 1033–1042. [CrossRef] [PubMed]
25. Du, Y.J.; Wei, M.L.; Reddy, K.R.; Liu, Z.P.; Jin, F. Effect of acid rain ph on leaching behavior of cement stabilized lead-contaminated soil. *J. Hazard. Mater.* **2014**, *271*, 131–140. [CrossRef]
26. Qu, Q.L.; Deng, Z.L.; Yang, Y.B.; Li, Z.G.; Yang, B.L. A Manufacturing Method for Producing Composite Phosphorus Slag Powder. China Patent ZL2012102079216, 21 June 2012. (In Chinese).
27. Li, J.S.; Xue, Q.; Wang, P.; Li, Z.Z.; Liu, L. Effect of drying-wetting cycles on leaching behavior of cement solidified lead-contaminated soil. *Chemosphere* **2014**, *117*, 10–13. [CrossRef]
28. Li, K.G.; Zheng, D.P.; Huang, W.H. Mechanical behavior of sandstone and its neural network simulation of constitutive model considering cyclic drying-wetting effect. *Rock. Soil. Mech.* **2013**, *34*, 168–173.
29. Abdelhadi, K.; Latifa, O.; Khadija, B.; Lahcen, B. Valorization of mining waste and tailings through paste backfilling solution, Imiter operation, Morocco. *Int. J. Min. Sci. Tech.* **2016**, *26*, 511–516.
30. Zhou, K.P.; Liu, T.Y.; Hu, Z.X. Exploration of damage evolution in marble due to lateral unloading using nuclear magnetic resonance. *Eng. Geol.* **2018**, *244*, 75–85. [CrossRef]
31. Nehdi, M.L.; Suleiman, A.R.; Soliman, A.M. Investigation of concrete exposed to dual sulfate attack. *Cement Concrete Res.* **2014**, *64*, 42–53. [CrossRef]
32. Collepardi, M. A state-of-the-art review on delayed ettringite attack on concrete. *Cement Concrete Res.* **2003**, *25*, 401–407. [CrossRef]
33. Tixier, R.; Mobasher, B. Modeling of damage in cement-based materials subjected to external sulfate attack. I: Formulation. *J. Mater. Civil Eng.* **2003**, *15*, 305–313. [CrossRef]
34. Andrés, E.I.; Carlos, M.L.; Ignacio, C. Chemo-mechanical analysis of concrete cracking and degradation due to external sulfate attack: A meso-scale model. *Cement Concrete Comp.* **2011**, *33*, 411–423.
35. Benzaazoua, M.; Belem, T.; Bussière, B. Chemical factors that influence the performance of mine sulphidic paste backfill. *Cement Concrete Res.* **2002**, *32*, 1133–1144. [CrossRef]
36. Cefis, N.; Comi, C. Chemo-mechanical modelling of the external sulfate attack in concrete. *Cement Concrete Res.* **2017**, *93*, 57–70. [CrossRef]
37. Hadigheh, S.A.; Gravina, R.J.; Smith, S.T. Effect of acid attack on FRP-to-concrete bonded interfaces. *Constr. Build. Mater.* **2017**, *152*, 285–303. [CrossRef]
38. Andre, B.; Moien, R.; Tilo, P.; Cyrill, G.; Florian, S.; Marlene, S.; Claudia, B.; Isabel, G.; Florian, E.; Florian, M. Effect of very high limestone content and quality on the sulfate resistance of blended cements. *Constr. Build. Mater.* **2018**, *188*, 1065–1076.

39. Wang, Y.; Yuan, Q.; Deng, D.; Ye, T.; Fang, L. Measuring the pore structure of cement asphalt mortar by nuclear magnetic resonance. *Constr. Build. Mater.* **2017**, *137*, 450–458. [CrossRef]
40. Zhang, J.; Deng, H.W.; Taheri, A.; Deng, J.R.; Ke, B. Effects of Superplasticizer on the Hydration, Consistency, and Strength Development of Cemented Paste Backfill. *Minerals* **2018**, *8*, 381. [CrossRef]
41. Min, C.D.; Li, X.B.; He, S.Y.; Zhou, S.T.; Zhou, Y.N.; Yang, S.; Shi, Y. Effect of mixing time on the properties of phosphogypsum-based cemented backfill. *Constr. Build. Mater.* **2019**, *210*, 564–573. [CrossRef]
42. Fall, M.; Benzaazoua, M. Modeling the effect of sulphate on strength development of paste backfill and binder mixture optimization. *Cement Concrete Res.* **2005**, *35*, 301–314. [CrossRef]
43. Wu, J.Y.; Feng, M.M.; Chen, Z.Q.; Mao, X.B.; Han, G.S.; Wang, Y.M. Particle Size Distribution Effects on the Strength Characteristic of Cemented Paste Backfill. *Minerals* **2018**, *8*, 322. [CrossRef]
44. Daisuke, S. Chemical alteration of calcium silicate hydrate (C–S–H) in sodium chloride solution. *Cement Concrete Res.* **2008**, *38*, 1270–1275.
45. Zhou, X.S.; Lin, X.; Huo, M.J.; Zhang, Y. The hydration of saline oil-well cement. *Cement Concrete Res.* **1996**, *26*, 1753–1759. [CrossRef]
46. Komljenović, M.M.; Baščarević, Z.; Marjanović, N.; Nikolić, V. Decalcification resistance of alkali-activated slag. *J. Hazard Mater.* **2012**, *233–234*, 112–121. [CrossRef]
47. Najjar, M.F.; Nehdi, M.L.; Soliman, A.M.; Azabi, T.M. Damage mechanisms of two-stage concrete exposed to chemical and physical sulfate attack. *Constr. Build. Mater.* **2017**, *137*, 141–152. [CrossRef]
48. Tang, Z.; Li, W.G.; Ke, G.J.; Zhou, J.L.; Tam, V.W.Y. Sulfate attack resistance of sustainable concrete incorporating various industrial solid wastes. *J. Clean. Prod.* **2019**, *218*, 810–822. [CrossRef]
49. Fan, Y.F.; Hu, Z.Q.; Zhang, Y.Z.; Liu, J.L. Deterioration of compressive property of concrete under simulated acid rain environment. *Constr. Build. Mater.* **2010**, *24*, 1975–1983. [CrossRef]
50. Brackebusch, F.W. Basics of paste backfill systems. *Miner. Eng.* **1994**, *46*, 1175–1178.
51. Zhao, Y.; Soltani, A.; Taheri, A.; Karakus, M.; Deng, A. Application of Slag–Cement and Fly Ash for Strength Development in Cemented Paste Backfills. *Minerals* **2019**, *9*, 22. [CrossRef]
52. Geng, J.; Easterbrook, D.; Li, L.Y.; Mo, L.W. The stability of bound chlorides in cement paste with sulfate attack. *Cement Concrete Res.* **2015**, *68*, 211–222. [CrossRef]
53. Al-Masri, M.S.; Amin, Y.; Ibrahim, S.; Al-Bich, F. Distribution of some trace metals in Syrian phosphogypsum. *Appl. Geochem.* **2004**, *19*, 747–753. [CrossRef]
54. Li, X.B.; Zhou, Y.N.; Zhu, Q.Q.; Zhou, S.T.; Min, C.D.; Shi, Y. Slurry Preparation Effects on the Cemented Phosphogypsum Backfill through an Orthogonal Experiment. *Minerals* **2019**, *9*, 31. [CrossRef]

 © 2019 by the authors. Licensee MDPI, Basel, Switzerland. This article is an open access article distributed under the terms and conditions of the Creative Commons Attribution (CC BY) license (http://creativecommons.org/licenses/by/4.0/).

Article

Radioactivity of Five Typical General Industrial Solid Wastes and its Influence in Solid Waste Recycling

Zhihui Shen [1,2,3,4], Qin Zhang [2,3,4,*], Wei Cheng [2,3,4] and Qianlin Chen [2,3,4,5]

1. College of Resources and Environmental Engineering, Guizhou University, Guiyang 550025, China; zhshen@gzu.edu.cn
2. College of Mining, Guizhou University, Guiyang 550025, China; wcheng1@gzu.edu.cn (W.C.); qlchen@gzu.edu.cn (Q.C.)
3. National & Local Joint Laboratory of Engineering for Effective Utilization of Regional Mineral Resources from Karst Areas, Guiyang 550025, China
4. Guizhou Key Laboratory of Comprehensive Utilization of Non-metallic Mineral Resources, Guiyang 550025, China
5. School of Chemistry and Chemical Engineering, Guizhou University, Guiyang 550025, China
* Correspondence: zq6736@163.com

Received: 14 January 2019; Accepted: 5 March 2019; Published: 9 March 2019

Abstract: The level of radionuclides is an important index for the preparation of building materials from industrial solid waste. In order to investigate the radiological hazard of five kinds of typical general industrial solid wastes in Guizhou, China, including fly ash (FA), red mud (RM), phosphorus slag (PS), phosphogypsum (PG), and electrolytic manganese residue (EMR), the radiation intensity and associated radiological impact were studied. The results show that concentrations of ^{238}U, ^{235}U, ^{232}Th, ^{226}Ra, ^{210}Pb, and ^{40}K for different samples vary widely. The concentration of ^{238}U was both positively correlated with ^{235}U and ^{226}Ra, and the uranium contents in the measured samples were all of natural origin. The radiation levels of PG, EMR, EMR-Na (EMR activated by NaOH), and EMR-Ca (EMR activated by Ca(OH)$_2$) were all lower than the Chinese and the world's recommended highest levels for materials allowed to be directly used as building materials. The values of the internal and external illumination index (I_{Ra} and I_γ, respectively) for FA and RM were higher ($I_{Ra} > 1.0$ and $I_\gamma > 1.3$ for FA, $I_{Ra} > 2.0$ and $I_\gamma > 2.0$ for RM). The radium equivalent activity (Ra$_{eq}$), indoor and outdoor absorbed dose (D_{in} and D_{out}, respectively), and corresponding annual effective dose rate (E_{in} and E_{out}) of RM, PS, and FA were higher than the recommended limit values (i.e., 370 Bq/kg, 84 nGy/h, 59 nGy/h, 0.4 mSv/y, and 0.07 mSv/y, respectively), resulting from the higher relative contribution of ^{226}Ra and ^{232}Th. The portion of RM, FA, and PS in building materials should be less than 75.44%, 29.72%, and 66.01%, respectively. This study provides quantitative analysis for the safe utilization of FA, RM, PS, PG, and EMR in Guizhou building materials.

Keywords: general industrial solid waste; building materials; natural radioactivity; activity concentration; radiological influence

1. Introduction

General industrial solid waste refers to waste discharged in various industrial production processes, such as the waste of the electric power industry, aluminum industry, phosphorus chemical industry, coal chemical industry, metallurgy, and so forth. This includes fly ash (FA), red mud (RM), phosphogypsum (PG), phosphorus slag (PS), and electrolytic manganese residue (EMR). At present, most general industrial solid waste is mainly used as raw materials for construction, such as cement and concrete, as well as environmental functional materials and a source of valuable elements [1]. In 2016, the comprehensive utilization of general industrial solid waste in China accounted for 48.0% of the country's total utilization and disposal, and disposal and storage accounted for 21.2% and 30.7% [2], respectively. As mineral resources usually contain natural radionuclides, the activity of radionuclides, such as FA, PS, PG, RM, and EMR, in some industrial solid wastes tends to increase after roasting or electrolyzation. When those general industrial solid wastes are used to prepare building materials, the gamma rays released by radionuclides can pose a radiation hazard to the living environment.

Natural radionuclides, ubiquitously spread in the natural environment [3,4], consist of three well-known radioactive series, that is, the actinium series originating from ^{235}U, the uranium series originating from ^{238}U, and the thorium series originating from ^{232}Th. In addition, there are several isolated radionuclides, such as ^{40}K [5,6]. The radioactive decay chains of ^{232}Th, ^{238}U, ^{235}U, and ^{40}K are the main contributors to the dose of natural radiation [7]. When the ratio of ^{235}U/^{238}U is less than 1%, the contribution of ^{235}U to the environmental dose is very small [8]. Since 98.5% of the radiological effects from uranium series nuclides are produced by ^{226}Ra and its decay products, the radiation from ^{238}U and the other ^{226}Ra precursors are negligible [9]. When industrial solid waste is used to prepare building materials, radionuclides in the environment can be determined by the natural radioactivity level of building materials or the industrial solid waste. Thus, radiation hazards can be assessed. At present, the internal and external exposure index, radium equivalent activity, and indoor and outdoor absorbed dose rate are commonly used indicators in evaluating radionuclide radiation hazards.

The average activity concentrations of ^{226}Ra, ^{232}Th, and ^{40}K in FA, RM, PG, and EMR samples from similar studies in different parts of the world are illustrated in Table 1. As shown in Table 1, it is suggested that the mean activity concentrations of ^{226}Ra for FA in Turkey, Greece, and China (Xiangyang) are 360, 815, and 441 Bq/kg, respectively, higher than those in other countries. Concentrations of ^{40}K in EU countries are generally higher than in others. In terms of RM, the average activity concentrations of ^{226}Ra, ^{232}Th, and ^{40}K in Australia and Jamaica are notably higher than those in other countries. In addition, ^{226}Ra is the main nuclide for PG in Israel, Spain, Korea, Egypt, and Turkey, while the concentrations of ^{226}Ra, ^{232}Th, and ^{40}K for PG in South Africa are relatively balanced. It can be concluded from Table 1 that radionuclide activity concentrations differ from one location to another.

Table 1. The obtained average activity concentrations of ^{226}Ra, ^{232}Th, and ^{40}K in FA, RM, PG, and EMR samples from similar studies in different parts of the world (unit: Bq/kg).

Country	Sample Type	Activity Concentration (Bq/kg)			Reference
		^{226}Ra	^{232}Th	^{40}K	
India		119	147	352	[10]
Turkey		360	102	517	[11]
Hungary		178	55	387	[12]
Greece		815	56	400	[13]
Czech Rep.		146	86	669	[13]
Germany	FA	164	94	517	[13]
Italy		170	140	400	[13]
Poland		200	118	798	[13]
Romania		219	116	595	[13]
China (Baoji)		112	148	386	[14]
China (Xiangyang)		441	110	510	[15]
Turkey		210	539	112	[13]
Hungary		301	295	50	[13]
Greece		244	364	57	[13]
Germany	RM	171	318	215	[13]
Italy		97	118	115	[13]
Australia		310	1350	350	[16]
Jamaica		1047	350	335	[17]

Table 1. Cont.

Country	Sample Type	Activity Concentration (Bq/kg)			Reference
		^{226}Ra	^{232}Th	^{40}K	
Turkey	PG	436	9	13	[18]
Bangladesh		234	21	108	[19]
Egypt		596	6	2	[20]
Jordan		376	4	40	[21]
Korea		618	9	24	[22]
Israel		747	14	63	[23]
Spain		647	8	33	[24]
South Africa		109	189	>100	[25]
Hungary	EMR	52	40	607	[26]
China (Chongqing)		37	58	631	[27]

Additionally, many studies have analyzed the radiation hazards of industrial solid wastes. Mamta Gupta et al. [10] reported that the radium equivalent activity of the FA from a thermal power plant in India was 353.9 Bq/kg, which is close to the maximum upper limit of 370 Bq/kg. L. Taoufiq et al. [28] characterized the radioactivity of FA from a thermal power plant in Morocco, and found that the radium equivalent activity was 241–298 Bq/kg, which is lower than 370 Bq/kg. In summary, the radionuclide activity concentrations and associated radiation hazards differ from one location to another. Different isotopes concentration in ores is a result of different conditions, including the metallogenetic body, formation age, epigenetic transformation, and so forth, during deposit formation in different regions. Therefore, the radioactive level of general industrial solid wastes in other regions cannot be used as the reference for Guizhou, China.

The dominant mineral resources, such as coal, phosphorite, bauxite, and manganese, are found in Guizhou. During the development and utilization process for these resources, a large amount of general solid waste including FA, PS, PG, RM, and EMR is accumulated. According to the data from the China statistical yearbook as shown in Table 2, from 2012 to 2017, the national generation volume of general industrial solid wastes varied from 3.09 to 3.32 billion tons, while the generation of Guizhou province varied from 0.071 to 0.094 billion tons. The substantial discharge and stockpiling of those aforementioned industrial solid wastes result in serious environmental pollution. It is vital to find an optimal solution for applying the solid wastes. The study of natural radionuclides and their radiation hazards is of great significance for the comprehensive utilization of industrial solid waste resources in the field of building materials.

Table 2. Generation of general industrial solid waste during 2012–2017, in Guizhou, China (unit: billion tons).

Year	China	Guizhou	Reference
2012	3.290	0.078	[29]
2013	3.277	0.082	[30]
2014	3.256	0.074	[31]
2015	3.271	0.071	[32]
2016	3.092	0.078	[33]
2017	3.316	0.094	[34]

In this study, the radionuclide activity of five typical general industrial solid wastes including FA, PS, RM, PG, and EMR in Guizhou, China, was measured using a gamma spectrometry technique. The radioactivity level and associated radiation hazard of these industrial solid wastes were evaluated using indicators such as the internal and external irradiation index (I_{Ra} and I_γ, respectively), radium equivalent ratio (Ra_{eq}), indoor and outdoor external dose (D_{in} and D_{out}, respectively), and indoor and outdoor annual effective dose rate (E_{in} and E_{out}, respectively). The maximum dosages of solid wastes in building materials were calculated. This study, through radiation evaluation, hopes to provide a mixing amount reference for the aforementioned solid wastes for building materials that meet the radiation limitation requirements.

2. Materials and Methods

2.1. Samples

The samples in this study include: PS, FA, RM, PG, EMR, electrolytic manganese slag activated by NaOH (EMR-Na), and electrolytic manganese slag activated by Ca(OH)$_2$ (EMR-Ca). Sample PS taken from a building material company in Guizhou, China, was produced using the electric furnace process of preparing yellow phosphorus, which is grayish white or white and partially agglomerated. Sample FA obtained from a coal-burning power plant in Guizhou was black. Sample RM produced by Bayer process from an aluminum company in Guizhou was light red. Sample PG obtained from a phosphorus chemical company in Guizhou was gray and very agglomerated. Sample EMR obtained from an electrolytic manganese plant in Guizhou was a black, fresh slurry. EMR has a certain potential gelling activity with a small amount of silicon and aluminum. The alkaline substance can better excite the potential activity and form a hydrated silicate and aluminate product with hydration characteristics, resulting in the gelling properties of EMR. The alkali-activated EMR can replace part of the cement used to prepare building materials. Previous studies showed that NaOH and Ca(OH)$_2$ have better activation effects on EMR. Therefore, EMR-Na used in this study was obtained by mixing 75 g of fresh EMR, 15 wt.% NaOH (accounting for EMR), and 100 mL of tap water in a 500 mL stirred tank and stirring this mixture for 20 min. EMR-Ca was obtained by mixing 100 g of fresh EMR, 20 wt.% Ca(OH)$_2$ (accounting for EMR), and 100 mL of tap water in a 500 mL stirred tank and stirring this mixture for 15 min.

The moisture contents of PS, FA, RM, PG, and EMR were 4.29%, 4.13%, 5.01%, 20.73%, and 15.7%, respectively. The densities of PS, FA, RM, PG, EMR, EMR-Na, and EMR-Ca were 3.06, 2.33, 3.01, 2.57, 2.31, 2.35, and 2.48 g/cm^3, respectively. Major components in seven solid wastes were analyzed by using X-ray fluorescence (XRF, PANalytical Axios mA×4 KW, Malvern Panalytical, Almelo, The Netherlands), and the results are illustrated in Table 3. As shown in Table 3, PS, RM, FA are mainly composed of CaO, SiO$_2$, Al$_2$O$_3$, and Fe$_2$O$_3$ [35], PG is composed of CaO, SiO$_2$, Al$_2$O$_3$, and SO$_3$ [36], while CaO, SiO$_2$, Al$_2$O$_3$, Mn, and S are the main components in EMR, EMR-Na, and EMR-Ca.

Table 3. Major components in seven solid wastes.

Component	Concentration (%)						
	PS	RM	FA	PG	EMR	EMR-Na	EMR-Ca
CaO	44.77	14.35	2.83	34.07	7.32	9.1	15.43
Fe$_2$O$_3$	0.45	21.53	2.83	0.20	-	-	-
Al$_2$O$_3$	5.55	20.89	14.75	0.13	8.85	11.27	7.99
SiO$_2$	37.81	16.75	45.72	5.29	22.85	30.29	20.35
MgO	2.61	1.55	1.18	0.01	1.86	2.42	1.80
P$_2$O$_5$	2.86	0.31	-	0.75	-	-	-
TiO$_2$	-	4.59	1.74	-	0.25	0.33	0.21
K$_2$O	1.43	0.98	1.24	-	1.75	2.22	1.66
Na$_2$O	0.33	4.93	0.55	-	0.14	0.42	0.12
SO$_3$	-	1.19	-	40.24	-	-	-
Mn	-	-	-	-	4.92	5.57	6.25
S	-	-	-	-	8.19	3.2	7.08
Others	0.04	12.04	17.03	19.31	2.88	3.75	2.18

2.2. Methods

2.2.1. Radioactivity Measurement

All samples were aggregated, identified, and oven-dried to constant weight in the laboratory, then grounded to a particle size of less than 0.075 mm. Each sample was homogenized and dried in an oven at 105 °C for 3 h to remove moisture. Then, 200 g of each sample were weighed and placed into a cylinder measuring sample box with a diameter of 35 mm and a height of 20 mm, then sealed for eight weeks to achieve radioactive secular equilibrium between ^{226}Ra and its daughters. The measurements of activity concentrations were carried out at the Institute of Geochemistry, Chinese Academy of Sciences, using a vertical closed coaxial HPGe detector (GX6020, CANBERRA, Oak Ridge, TN, USA). The detector has an energy range of 3 keV to 10 MeV, with an energy resolution of 2.0 keV. The relative efficiency is 60% at 1332 keV γ-ray, and the peak-to-Compton ratio is 66:1. The actual energy range used to test the samples was 35–3000 keV. The measurement time for each sample was set as 180,000 s. The test data was collated and analyzed using management software OpenEMS with a data management system (RDBMS). To reduce the gamma ray background, a cylindrical lead shield detector was used to absorb X-rays, which contains two inner concentric cylinders of copper and aluminum. The calibration sources with an energy range covering nuclides ^{238}U, ^{226}Ra, ^{232}Th, ^{40}K, ^{241}Am, ^{137}Cs, and ^{60}Co were used for the determination of the detection efficiency of the measurement system, and the typically obtained values were within 6% of certified values.

The γ-ray lines that were used to measure the activities for nuclides were represented mainly by gamma-ray-emitting nuclei in the decay series of ^{232}Th, ^{226}Ra, and ^{40}K. The ^{40}K and ^{210}Pb were determined from their single photo peaks of 1460 keV (1.2%) and 46.5 keV (4.26%), respectively. The ^{238}U and ^{232}Th are not gamma ray emitters. However, it is possible to measure the gamma rays of their decay products. The decay product taken for ^{238}U was ^{234}Th (63.3 keV (3.81%)). The intensity of gamma rays emitted by ^{232}Th is very weak, and its decay product ^{228}Ra has a long half-life without gamma rays. The half-life of ^{228}Ac, the daughter of ^{228}Ra, is 6.13 h. According to the literature, Th and Ra both have affinity for silicates [37]. Moreover, some studies have shown that the activities of ^{232}Th and ^{228}Ra in solid wastes such as red mud, phosphogypsum, and fly ash are approximately balanced [38–40]. Therefore, the equilibrium between ^{232}Th and ^{228}Ra was assumed, and the ^{232}Th was determined using ^{208}Tl (583.2 keV (30.78%) and 2614.5 keV (35.7%)) and ^{228}Ac (911.2 keV (26.6%)). In addition, the ^{214}Pb (351.9 keV (35.8%)) and ^{214}Bi (609.3 keV (45%)) were used to determine the ^{226}Ra, and the decay product taken for ^{235}U was ^{235}U (185.7 keV (57.5%)). As there is an interference between 185.7 keV from ^{235}U and 186.2 keV from ^{226}Ra, the ^{235}U activity can be deduced from the 186 keV multiplet after removal of the ^{226}Ra contribution. Thus, the activity concentration of ^{235}U radionuclide is given by Equation (1) [41]:

$$A_{U-235} = 1.75 \times \frac{CR_{total(186\ keV)}}{\varepsilon_{186\ keV}} - 0.06 A_{Ra-226} \qquad (1)$$

where CR and $CR_{total(186\ keV)}$ are calculated by using Equations (2) and (3):

$$CR = A \times BR \times \varepsilon \qquad (2)$$

$$CR_{total(186\ keV)} = CR_{Ra-226} + CR_{U-235} \qquad (3)$$

where CR is the counting rate in full-energy peak in count/s, $CR_{total(186\ keV)}$ is the counting rate for the 186 keV multiple, A is the activity of radionuclide in Bq/kg, BR is the branching ratio or the gamma-ray emission rate, ε is the detection efficiency. ^{226}Ra and ^{235}U contribute about 58% and 42% of the total count rate of the 186 keV peak with the existence of equilibrium, respectively [42].

2.2.2. Internal and External Illumination Index

The internal exposure index (I_{Ra}) refers to the specific activity ratio of ^{226}Ra in the building materials to the ^{226}Ra limit specified in the national standard (GB6566-2010) [43]. The external radiation index (I_γ) refers to the sum of the specific activity ratio of ^{226}Ra, ^{232}Th, and ^{40}K in building materials to their respective standard limits. The I_{Ra} and I_γ were calculated by using Equations (4) and (5):

$$I_{Ra} = \frac{C_{Ra}}{200} \tag{4}$$

$$I_\gamma = \frac{C_{Ra}}{370} + \frac{C_{Th}}{260} + \frac{C_K}{4200} \tag{5}$$

where C_{Ra}, C_{Th}, and C_K are the mean radioactivity concentrations of ^{226}Ra, ^{232}Th, and ^{40}K (Bq/kg), respectively. The specific activity limit of ^{226}Ra in building materials specified in the GB6566-2010 is 200 Bq/kg, considering only the internal irradiation conditions. The prescribed limits were 370, 260, and 4200 Bq/kg for ^{226}Ra, ^{232}Th, and ^{40}K, respectively, in building materials when they exist separately under the external irradiation condition.

2.2.3. Radium Equivalent Activity

The distribution of ^{226}Ra, ^{232}Th, and ^{40}K in building materials is not uniform [44]. The radium equivalent activity (Ra_{eq}) [45] was used to compare the relative gamma radioactivity of ^{226}Ra, ^{232}Th, and ^{40}K in building materials. It has been estimated that 1 Bq/kg of ^{226}Ra, 0.7 Bq/kg of ^{232}Th, and 13 Bq/kg of ^{40}K produce the same gamma ray dose [46,47]. Thus, the radium equivalent activities (Ra_{eq}) can be estimated using Equation (6) [48–50]:

$$Ra_{eq} = C_{Ra} + 1.43C_{Th} + 0.077C_K \tag{6}$$

2.2.4. Indoor External Dose (D_{in}) and Outdoor External Dose (D_{out})

An irradiation scenario is required to evaluate the ^{226}Ra, ^{232}Th, and ^{40}K absorbed doses produced by the building materials. The European Commission proposed that the length, width, and height of the concrete room are 4, 5, and 2.8 m, respectively. The thickness and density of the wall are 20 cm and 2350 kg·m^{-3}, respectively [51]. Then, the indoor external dose D_{in} could be calculated by using Equation (7):

$$D_{in}(nGy/h) = 0.92C_{Ra} + 1.1C_{Th} + 0.081C_K. \tag{7}$$

To assess the dose of radiation from building materials in a room, the portion from natural radiation needs to be subtracted. According to the survey results of the Chinese National Environmental Protection Department, the weighted mean by area and population are 62.8 and 62.1 nGy·h^{-1}, respectively. Taking 62.1 nGy·h^{-1} as the natural radiation background value, the calculated absorbed dose rate should be reduced by 62.1 [52].

The outdoor external dose (D_{out}) assessed from ^{226}Ra, ^{232}Th, and ^{40}K was supposed to be equally distributed at 1 m above the ground surface. Therefore, the D_{out} was calculated using Equation (8) [53]:

$$D_{out}(nGy/h) = 0.462C_{Ra} + 0.604C_{Th} + 0.0417C_K. \tag{8}$$

2.2.5. Annual Effective Dose Rate

Buildings are the main places for daily activities of human beings, and the indoor and outdoor occupancy factors are 0.8 and 0.2, respectively (i.e., 80% and 20% of the time they are occupied indoors and outdoors, respectively) [54,55]. The conversion factor from the absorbed dose in the air to the effective dose received by the individual is 0.7 [56]. The annual indoor effective dose rate

(E_{in}) and annual outdoor effective dose rate (E_{out}) can be calculated using Equations (9) and (10), respectively [55,57]:

$$E_{in} = D_{in} \times 365 \times 24\,h \times 0.8 \times 0.7 (Sv/Gy) = D_{in} \times 4905 \times 10^{-6} (mSv/y) \qquad (9)$$

$$E_{out} = D_{out} \times 365 \times 24\,h \times 0.2 \times 0.7 (Sv/Gy) = D_{out} \times 1226 \times 10^{-6} (mSv/y) \qquad (10)$$

2.2.6. Maximum Dosage of Solid Waste in Building Materials

The maximum dosage of solid waste in building materials f_s can be calculated by using Equations (11) and (12):

$$\frac{f_s \cdot C_{Ra} + (1-f_s) \cdot C'_{Ra}}{200} \leq 1.0 \qquad (11)$$

$$\frac{f_s \cdot C_{Ra} + (1-f_s) \cdot C'_{Ra}}{370} + \frac{f_s \cdot C_{Th} + (1-f_s) \cdot C'_{Th}}{260} + \frac{f_s \cdot C_K + (1-f_s) \cdot C'_K}{4200} \leq 1.0 \qquad (12)$$

where C'_{Ra}, C'_{Th}, and C'_K are the mean radioactivity concentrations of ^{226}Ra, ^{232}Th, and ^{40}K (in Bq/kg) for other components in building materials, respectively. When the ratio of C'_{Ra}, C'_{Th}, and C'_K are all meant to be zero, the ratio of f_s calculated using Equations (11) and (12) is the maximum dosage of solid waste in building materials.

3. Results and Discussion

3.1. Activity Concentration

The average values of the activity concentration of six nuclides for seven samples were calculated and are illustrated in Table 4. It can be concluded that the seven industrial solid wastes all contained ^{238}U, ^{235}U, ^{232}Th, ^{226}Ra, ^{210}Pb, and ^{40}K. The activity concentrations of ^{40}K were relatively high in EMR, EMR-Na, PS, and FA, with values of 443.8, 423.9, 529.4, and 461.0 Bq/kg, respectively. In FA, RM, and PS, the activity concentrations of ^{226}Ra were relatively higher than those of the other samples, at 208.2, 462.7, and 187.4 Bq/kg, respectively. The activity concentrations of ^{232}Th in FA and RM were as high as 165.6 and 457.7 Bq/kg, respectively. In addition, activity concentrations of ^{238}U in FA, RM, and PS were higher than those in other solid wastes, which were 234.9, 513.0, and 199.8 Bq/kg, respectively. On the contrary, the contents of ^{210}Pb and ^{235}U were very low in all samples. Furthermore, when NaOH and Ca(OH)$_2$ were used to activate EMR, the activity concentrations of the six nuclides in EMR-Na and EMR-Ca decreased, indicating that the addition of NaOH and Ca(OH)$_2$ can weaken the radioactivity of EMR.

Table 4. The activity concentrations (Bq/kg) of six nuclides for the seven solid wastes.

Name of Solid Wastes	Activity Concentration (Bq/kg)					
	^{40}K	^{210}Pb	^{226}Ra	^{232}Th	^{235}U	^{238}U
PS	461.0 ± 23.6	15.1 ± 9.9	187.4 ± 5.8	233.7 ± 9.6	9.1 ± 0.3	199.8 ± 17.9
RM	259.5 ± 18.9	324.8 ± 18.1	462.7 ± 14.5	457.7 ± 13.4	22.4 ± 0.7	513.0 ± 43.1
FA	529.4 ± 31.1	210.5 ± 11.8	208.2 ± 7.0	165.6 ± 5.6	10.1 ± 0.3	234.9 ± 27.2
PG	3.3 ± 6.0	86.0 ± 4.8	61.0 ± 2.1	2.3 ± 0.5	2.7 ± 0.1	28.0 ± 13.0
EMR	443.8 ± 26.7	34.3 ± 2.3	26.6 ± 1.6	24.8 ± 1.7	1.6 ± 0.1	48.3 ± 15.9
EMR-Na	423.9 ± 25.9	45.9 ± 2.9	24.5 ± 1.6	22.2 ± 1.5	1.5 ± 0.1	37.9 ± 15.7
EMR-Ca	321.4 ± 20.9	37.7 ± 2.5	21.1 ± 1.5	20.9 ± 1.6	1.3 ± 0.1	39.8 ± 15.8

3.2. The Source Analysis of Uranium

The analysis of the relationship between ^{238}U and ^{235}U and ^{226}Ra can be used to trace the source of the radioactive contamination by uranium in the environment. The activity concentrations of ^{238}U were plotted against the activity concentrations of ^{235}U and ^{226}Ra, as shown in Figure 1. The ratios of ^{238}U/^{235}U and ^{238}U/^{226}Ra were calculated and are shown in Table 4. As shown in Figure 1a, the linear fitting of the graph shows a good correlation between ^{238}U and ^{235}U ($R^2 = 0.990$). From the natural isotope abundance of uranium isotopes (^{238}U is 99.2%, ^{235}U is 0.72%) and its half-life, it is well known that naturally occurring uranium has a constant ^{238}U/^{235}U activity ratio of 21.7 [58]. The results given in Table 4 indicate that the ratios of ^{238}U/^{235}U for the seven samples vary from 10.37 to 30.62, and the ^{238}U/^{235}U of PG is the lowest at 10.37. The ratio of PS, RM, and FA is close to the constant value, while the ^{238}U/^{235}U of EMR, EMR-Na, and EMR-Ca are all higher than 21.7. The reason for the deviation may be the higher uncertainty values caused by the self-absorption effect. Additionally, the emanation of radon from the sealed samples may also cause an underestimation of uranium activity concentrations [59]. As CaO, SiO$_2$, Al$_2$O$_3$ are the major components in PS, RM, FA, and PG, according to the literature, the accumulations of U are often associated with clays because the clay fraction has an affinity for absorbing U; ^{238}U may be easily enriched in clay minerals from the perspective of adsorption [60]. PS, RM, FA, and PG are obtained in the process of calcination, coal alumina, dissolution combustion, and phosphoric acid production, thus the adsorption of clay minerals may have little effect on uranium migration.

As shown in Table 5, the ratios of ^{238}U/^{226}Ra for the seven solid wastes varied from 0.46 to 1.88, and the concentrations of ^{238}U were commonly lower than ^{226}Ra. The reason for this may be that ^{238}U/^{226}Ra was disturbed in favor of ^{226}Ra [59]. As shown in Figure 1b, the linear fitting of the graph shows a good correlation ($R^2 = 0.985$), and the slope of the line has a value (1.09) close to the average value of 1.29 for ^{238}U/^{226}Ra activity ratios. This indicates that there may exist a radioactive balance between ^{238}U and ^{226}Ra. According to previous literature reports, there is depleted uranium pollution in addition to natural uranium in the sample when the ^{238}U/^{226}Ra ratio is greater than 5 [61]. Based on this information, it may be concluded that the uranium contents in the measured seven samples are all of natural origin.

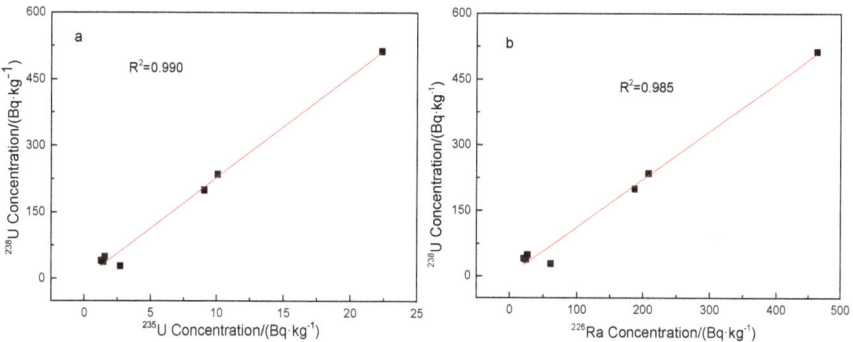

Figure 1. Variation of ^{238}U activity concentration versus (**a**) ^{235}U activity and (**b**) ^{226}Ra activity.

Table 5. The activity ratio for different samples under investigation.

Name of Solid Wastes	^{238}U/^{235}U	^{238}U/^{226}Ra
PS	21.96	1.07
RM	22.90	1.11
FA	23.26	1.13
PG	10.37	0.46
EMR	30.19	1.82
EMR-Na	25.27	1.54
EMR-Ca	30.62	1.88

3.3. Radiological Impact Assessment

3.3.1. Analysis of Radiation Hazard Indexes

The I_{Ra}, I_γ, Ra_{eq}, D_{in}, D_{out}, E_{in}, and E_{out} were calculated according to the activity concentrations of the radionuclides of the seven solid wastes and Equations (1)–(7). As illustrated in Table 6, the I_{Ra} and I_γ of the seven industrial solid wastes were 0.11–2.31 and 0.17–3.07 Bq/kg, respectively, of which PS, PG, EMR, EMR-Na, and EMR-Ca were all less than 1. The I_{Ra} was greater than 1 and the I_γ was greater than 1.3 for FA. The I_{Ra} and I_γ of the RM were both greater than 2. According to Table 7, "Limited Standards for Radionuclide of Building Materials", PS, PG, EMR, EMR-Na, and EMR-Ca can be directly used as building materials and decorative materials of class A, B, and C. FA can be used as decoration materials of class B and class C. This means that almost all samples are safe for use as they meet the PRC National Standard. Adding FA and RM in building materials should be considered.

According to Table 6, the Ra_{eq} values of the seven samples were between 64.54 and 1137.18 Bq/kg, among which the Ra_{eq} of PG, EMR, EMR-Na, and EMR-Ca were lower than the world's recommended limit (370 Bq/kg) for building materials [56]. The Ra_{eq} values of the RM, PS, and FA were as high as 1137.18, 557.03, and 485.78 Bq/kg, respectively. Therefore, the dosage of RM, PS, and FA should be considered when used as the source for building materials.

It also can be seen from Table 6 that the D_{in} and D_{out} of the seven industrial solid wastes were 6.38–888.07 and 29.70–501.03 nGy/h, respectively. Specifically, the D_{in} of PS, RM, and FA were as high as 404.67, 888.07, and 354.49 nGy/h, respectively, while their D_{out} values were 246.93, 501.03, and 218.29 nGy/h, respectively. These values are higher than the world's average values (i.e., 84 nGy/h for D_{in} and 59 nGy/h for D_{out} [56,62]). The results show that the values of the annual effective dose rate for the seven samples were 0.03–4.36 mSv/y for E_{in} and 0.04–0.61 mSv/y for E_{out}. The values of E_{in} and E_{out} for PS, RM, and FA were all higher than the world's recommended values (i.e., 0.4 mSv/y for E_{in} and 0.07 mSv/y for E_{out} [56]).

In summary, the radioactive levels of RM, FA, and PS exceed the "Limited Standards for the Radionuclide of Building Materials", and they cannot be directly used for building materials, while the PG, EMR, EMR-Na, and EMR-Ca could be directly used for building materials within the recommended levels.

Table 6. The I_{Ra}, I_γ, Ra_{eq}, D_{in}, D_{out}, E_{in}, and E_{out} for industrial solid wastes.

Solid Wastes	I_{Ra} (Bq/kg)	I_γ (Bq/kg)	Ra_{eq} (Bq/kg)	D_{in} (nGy/h)	D_{out} (nGy/h)	E_{in} (mSv/y)	E_{out} (mSv/y)
PS	0.94	0.71	557.03	404.67	246.93	1.98	0.30
RM	2.31	3.07	1137.18	888.07	501.03	4.36	0.61
FA	1.04	1.33	485.78	354.49	218.29	1.74	0.27
PG	0.30	0.17	64.54	−3.19 *	29.70	−0.02 *	0.04
EMR	0.13	0.27	96.23	25.60	45.77	0.13	0.06
EMR-Na	0.12	0.25	88.99	19.28	42.45	0.09	0.05
EMR-Ca	0.11	0.21	75.78	6.38	35.80	0.03	0.04

* The minus sign represents that the D_{in} of PG is lower than the natural radiation background value.

Table 7. Limited Standards for the Radionuclide of Building Materials (GB/6566-2010) [43].

Standard Name	Building Materials		Decoration Materials		
	Hollow Rate ≤ 25%	Hollow Rate > 25%	Class A	Class B	Class C
Radionuclide limit	$I_{Ra} \leq 1.0$ $I_r \leq 1.0$	$I_{Ra} \leq 1.0$ $I_r \leq 1.3$	$I_{Ra} \leq 1.0$ $I_r \leq 1.3$	$I_{Ra} \leq 1.3$ $I_r \leq 1.9$	$I_r \leq 2.8$

3.3.2. Contribution Analysis of Nuclides to Radiation

According to Equations (1)–(7), the contribution of radionuclides ^{26}Ra, ^{232}Th, and ^{40}K to radiation hazard indexes varies in different solid wastes. The contributions of ^{26}Ra, ^{232}Th, and ^{40}K to I_γ, Ra_{eq}, D_{in}, E_{in}, D_{out}, and E_{out} were calculated and plotted in Figure 2. As shown in Figure 2, ^{226}Ra was the main contributor to I_γ, Ra_{eq}, D_{in}, E_{in}, D_{out}, and E_{out} in RM, FA, and PG, and the relative contributions of ^{226}Ra were in the range of 40.69–44.80% for RM, 42.44–45.97% for FA, and 92.44–95.19% for PG. Similarly, ^{232}Th was the main contributor to I_γ, Ra_{eq}, D_{in}, E_{in}, D_{out}, and E_{out} in PS, RM, and FA, and the relative contributions of ^{232}Th varied from 55.07 to 59.99% for PS, 52.98 to 57.55% for RM, and 43.73 to 48.76% for FA. However, the relative contributions of ^{40}K to I_γ, Ra_{eq}, D_{in}, E_{in}, D_{out}, and E_{out} in EMR, EMR-Na, and EMR-Ca were in the range of 35.51–41.00%, 36.68–42.19%, and 32.66–38.02%, respectively. This data shows that ^{226}Ra and ^{232}Th were the main contributors to radiation hazard indexes in PS, RM, FA, and PG, while the contributions of ^{226}Ra, ^{232}Th, and ^{40}K in EMR, EMR-Na, and EMR-Ca were relatively balanced. However, the phase analysis of known nuclides in those aforementioned solid wastes has not been covered in this study, and future research should focus on the removal or decrease of the nuclides in different phases.

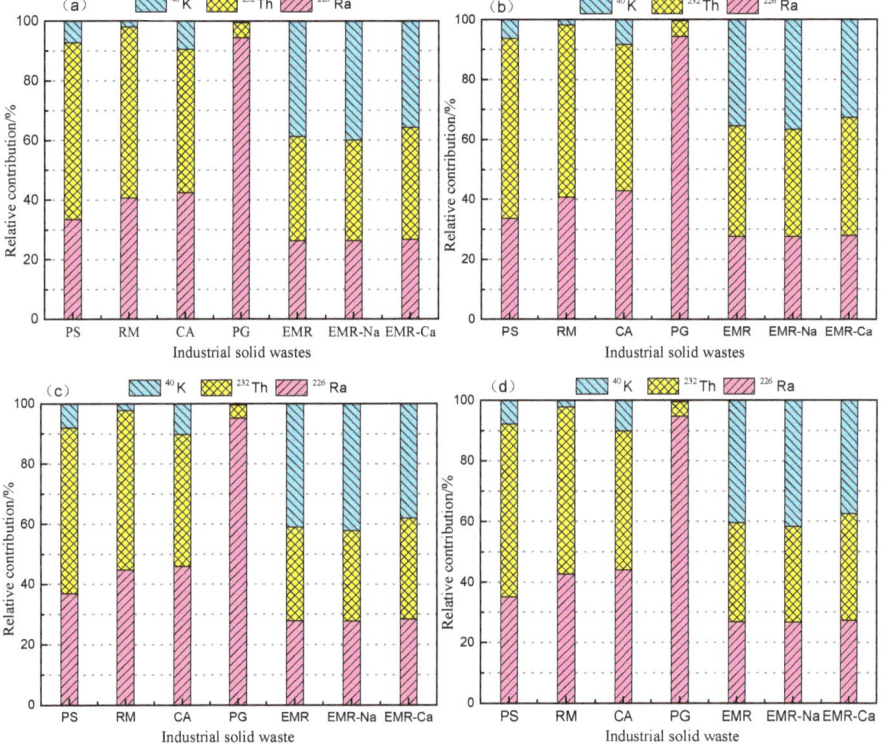

Figure 2. The relative concentrations of ^{226}Ra, ^{232}Th, and ^{40}K in (**a**) I_γ, (**b**) Ra_{eq}, (**c**) D_{in} and E_{in}, (**d**) D_{out} and E_{out} in industrial solid wastes.

3.3.3. Limitation Analysis of Solid Wastes in Building Materials

As the solid wastes RM, FA, and PS cannot be directly used for building materials, the maximum dosages (f_s) of solid wastes FA, RM, and PS in building materials were calculated using Equations (11) and (12) with results of 75.44%, 29.72%, and 66.01%, respectively. According to the different related research reviews related, increasing the addition of FA in concrete can result in a decrease in the compressive strength of concrete. The optimum FA percentages were found to be 10%, 20%, 22%, and 35% in various studies [63–66]. Similarly, previous research showed that the corrosion rate of concrete is the lowest between 20 wt. % and 30 wt. % of added RM content [67]. Other studies showed that the addition of RM in amounts greater than 20% causes a decrease in the hydration temperature and results in a decrease in the compressive strength [68]. Therefore, in terms of the mechanical properties of building materials, the optimum addition of FA, RM, and PS in building materials will not exceed the maximum addition allowed by their radioactivity level. That is, FA, RM, and PS can be used even if their radioactivity levels are above the standard limits.

4. Conclusions

(1) ^{40}K, ^{226}Ra, ^{232}Th, and ^{238}U are the main nuclides in FA, RM, and PS. ^{210}Pb and ^{226}Ra are the main nuclides in PG, while ^{40}K is the main nuclide in EMR, EMR-Na, and EMR-Ca. The uranium contents of all samples are all from natural uranium.

(2) The values of I_{Ra} and I_γ were all less than 1 except for FA ($I_{Ra} > 1$ and $I_\gamma > 1.3$) and RM ($I_{Ra} > 2$ and $I_\gamma > 2$), and the values of Ra_{eq}, D_{in}, D_{out}, E_{in}, and E_{out} were higher than the world's recommended values (i.e., 370 Bq/kg, 84 nGy/h, 59 nGy/h, 0.4 mSv/y, and 0.07 mSv/y, respectively) for PS, RM, and FA due to higher concentrations of ^{226}Ra and ^{232}Th.

(3) PG, EMR, EMR-Na, and EMR-Ca could be used for building materials unlimitedly. However, RM, FA, and PS could be used as additive or auxiliary materials for building materials by means of doping and mixing, with maximum portions of 75.44%, 29.72%, and 66.01%, respectively. These findings provide a basis for the restriction of the aforementioned solid wastes in Guizhou in the field of building materials. Further research on the phase analysis and treatment of known nuclides in solid wastes are necessary in the future.

Author Contributions: Z.S. and Q.Z. conceived and designed the experiments; Z.S. performed the experiments; Q.Z. contributed the analysis resource; W.C. and Q.C. helped to analyze the data; Z.S. wrote the paper.

Funding: This study was founded by the Science and Technology Program of Guizhou Province, China (Qian Ke He Support [2017]2040) and the Major Special Projects of Guizhou Province, China (Qian Ke He [20116023].

Conflicts of Interest: The authors declare no conflict of interest.

References

1. Zhang, Y.H.; Wang, X.K.; Lu, F.Z. Study progress of alkali removal from red mud and novel functional materials (in Chinese). *Chin. J. Environ. Eng.* **2016**, *10*, 3383–3390.
2. *2017 Annual Report on Prevention and Control of Solid Waste in China's Large and Medium-Sized Cities (In Chinese)*; Environmental Protection: Beijing, China, 2018; pp. 90–106.
3. Matiullah; Ahad, A.; Rehman, S.U.; Rehman, S.U.; Faheem, M. Measurement of radioactivity in the soil of Bahawalpur division, Pakistan. *Radiat. Prot. Dosim.* **2004**, *112*, 443. [CrossRef] [PubMed]
4. Mehra, R.; Singh, S.; Singh, K.; Sonkawade, R. ^{226}Ra, ^{232}Th and ^{40}K analysis in soil samples from some areas of Malwa region, Punjab, India using gamma ray spectrometry. *Environ. Monit. Assess.* **2007**, *134*, 333–342. [CrossRef] [PubMed]
5. Ahmad, N.; Matiullah; Khatibeh, A.J.A.H.; Ma'Ly, A.; Kenawy, M.A. Measurement of natural radioactivity in Jordanian sand. *Radiat. Meas.* **1997**, *28*, 341–344. [CrossRef]
6. Khatibeh, A.J.A.H.; Ahmad, N.; Matiullah; Kenawy, M.A. Natural radioactivity in marble stones- Jordan. *Radiat. Meas.* **1997**, *28*, 345–348. [CrossRef]
7. Bruzzi, L.; Baroni, M.; Mele, R.; Nanni, E. Proposal for a method of certification of natural radioactivity in building materials. *J. Radiol. Prot.* **1997**, *17*, 85. [CrossRef]
8. Chiozzi, P.; De, F.P.; Fazio, A.; Pasquale, V.V.; Verdoya, M. Laboratory application of NaI (Tl) gamma-ray spectrometry to studies of natural radioactivity in geophysics. *Appl. Radiat. Isot.* **2000**, *53*, 127–132. [CrossRef]
9. Turhan, S. Assessment of the natural radioactivity and radiological hazards in Turkish cement and its raw materials. *J. Environ. Radioact.* **2008**, *99*, 404–414. [CrossRef]

10. Gupta, M.; Mahur, A.K.; Varshney, R.; Sonkawade, R.G.; Verma, K.D.; Prasad, R. Measurement of natural radioactivity and radon exhalation rate in fly ash samples from a thermal power plant and estimation of radiation doses. *Radiat. Meas.* **2013**, *50*, 160–165. [CrossRef]
11. Turhan, Ş.; Parmaksız, A.; Köse, A.; Yüksel, A.; Arıkan, İ.H.; Yücel, B. Radiological characteristics of pulverized fly ashes produced in Turkish coal-burning thermal power plants. *Fuel* **2010**, *89*, 3892–3900. [CrossRef]
12. Kardos, R.; Sas, Z.; Shahrokhi, A.; Somlai, J.; Kovács, T. Radionuclide content of NORM by-products originating from the coal-fired power plant in Oroszlány (Hungary). *Radiat. Prot. Dosim.* **2015**, *167*, 266–269. [CrossRef] [PubMed]
13. Trevisi, R.; Leonardi, F.; Risica, S.; Nuccetelli, C. Updated database on natural radioactivity in building materials in Europe. *J. Environ. Radioact.* **2018**, *187*, 90–105. [CrossRef]
14. Lu, X.; Zhang, X. Radionuclide content and associated radiation hazards of building materials and by-products in Baoji, West China. *Radiat. Prot. Dosim.* **2007**, *128*, 471–476. [CrossRef] [PubMed]
15. Feng, T.T.; Lu, X.W. Natural radioactivity, radon exhalation rate and radiation dose of fly ash used as building materials in Xiangyang, China. *Indoor Built Environ.* **2016**, *25*, 626–634. [CrossRef]
16. Cooper, M.B.; Clarke, P.C.; Robertson, W.; Mcpharlin, I.R.; Jeffrey, R.C. An investigation of radionuclide uptake into food crops grown in soils treated with bauxite mining residues. *J. Radioanal. Nuclear Chem.* **1995**, *194*, 379–387. [CrossRef]
17. Pinnock, W.R. Measurements of radioactivity in Jamaican building materials and gamma dose equivalents in a prototype red mud house. *Health Phys.* **1991**, *61*, 647–651. [CrossRef] [PubMed]
18. Gezer, F.; Turhan, Ş.; Uğur, F.A.; Gören, E.; Kurt, M.Z.; Ufuktepe, Y. Natural radionuclide content of disposed phosphogypsum as TENORM produced from phosphorus fertilizer industry in Turkey. *Ann. Nuclear Energy* **2012**, *50*, 33–37. [CrossRef]
19. Alam, M.N.; Chowdhury, M.I.; Kamal, M.; Ghose, S.; Banu, H.; Chakraborty, D. Radioactivity in chemical fertilizers used in Bangladesh. *Appl. Radiat. Isot.* **1997**, *48*, 1165–1168. [CrossRef]
20. Mourad, N.M.; Sharshar, T.; Elnimr, T.; Mousa, M.A. Radioactivity and fluoride contamination derived from a phosphate fertilizer plant in Egypt. *Appl. Radiat. Isot.* **2009**, *67*, 1259–1268. [CrossRef]
21. Al-Jundi, J.; Al-Ahmad, N.; Shehadeh, H.; Afaneh, F.; Maghrabi, M.; Gerstmann, U.; Hā Llriegl, V.; Oeh, U. Investigations on the activity concentrations of ^{238}U, ^{226}Ra, ^{228}Ra, ^{210}Pb and ^{40}K in Jordan phosphogypsum and fertilizers. *Radiat. Prot. Dosim.* **2008**, *131*, 449–454. [CrossRef]
22. Song, M.H.; Chang, B.U.; Koh, S.M.; Kim, Y.J.; Kim, D.J.K.G. Overall natural radioactivity of a phosphate fertilizer industry in Korea. *Radioprotection* **2012**, *46*, S113–S118. [CrossRef]
23. Kovler, K.; Haquin, G.; Manasherov, V.; Ne'Eman, E.; Lavi, N. Natural radionuclides in building materials available in Israel. *Build. Environ.* **2002**, *37*, 531–537. [CrossRef]
24. Dueñas, C.; Fernández, M.C.; Cañete, S.; Pérez, M. Radiological impacts of natural radioactivity from phosphogypsum piles in Huelva (Spain). *Radiat. Meas.* **2010**, *45*, 242–246. [CrossRef]
25. Msila, X.; Labuschagne, F.; Barnard, W.; Billing, D.G. Radioactive nuclides in phosphogypsum from the lowveld region of South Africa. *S. Afr. J. Sci.* **2016**, *112*, 1–5. [CrossRef]
26. Kovács, T.; Shahrokhi, A.; Sas, Z.; Vigh, T.; Somlai, J. Radon exhalation study of manganese clay residue and usability in brick production. *J. Environ. Radioact.* **2017**, *168*, 15–20. [CrossRef] [PubMed]
27. Du, B.; Zhou, C.; Dan, Z.; Luan, Z.; Duan, N. Preparation and characteristics of steam-autoclaved bricks produced from electrolytic manganese solid waste. *Constr. Build. Mater.* **2014**, *50*, 291–299. [CrossRef]
28. Taoufiq, L.; Laamyem, A.; Boukhair, A.; Essediqi, E.; Monkade, M.; Zrabda, A. Radiological assessment of wastewater treatment processes based on the use of coal ashes as a filters. *J. Radiat. Res. Appl. Sci.* **2018**. [CrossRef]
29. NBSC. *2013 China Statistical Yearbook (in Chinese)*; China Statistics Press: Beijing, China, 2013; pp. 278–279. ISBN 9787503769634.
30. NBSC. *2014 China Statistical Yearbook (in Chinese)*; China Statistics Press: Beijing, China, 2014; pp. 227–228. ISBN 9787503772801.
31. NBSC. *2015 China Statistical Yearbook (in Chinese)*; China Statistics Press: Beijing, China, 2015; pp. 251–252. ISBN 9787503776380.
32. NBSC. *2016 China Statistical Yearbook (in Chinese)*; China Statistics Press: Beijing, China, 2016; pp. 247–248. ISBN 9787503779176.

33. NBSC. *2017 China Statistical Yearbook (in Chinese)*; China Statistics Press: Beijing, China, 2017; pp. 241–242. ISBN 978-7-5037-8253-4.
34. NBSC. *2018 China Statistical Yearbook (in Chinese)*; China Statistics Press: Beijing, China, 2018; pp. 250–251. ISBN 9787503785870.
35. Li, X.; Ye, J.; Liu, Z.; Qiu, Y.; Li, L.; Mao, S.; Wang, X.; Zhang, Q. Microwave digestion and alkali fusion assisted hydrothermal synthesis of zeolite from coal fly ash for enhanced adsorption of Cd(II) in aqueous solution. *J. Cent. South Univ.* **2018**, *25*, 9–20. [CrossRef]
36. Li, X.; Zhang, Q.; Ke, B.; Wang, X.; Li, L.; Li, X.; Mao, S. Insight into the effect of maleic acid on the preparation of α-hemihydrate gypsum from phosphogypsum in Na_2SO_4 solution. *J. Cryst. Growth* **2018**, *493*, 34–40. [CrossRef]
37. Croymans, T.; Vandael Schreurs, I.; Hult, M.; Marissens, G.; Lutter, G.; Stroh, H.; Schreurs, S.; Schroeyers, W. Variation of natural radionuclides in non-ferrous fayalite slags during a one-month production period. *J. Environ. Radioact.* **2017**, *172*, 63–73. [CrossRef]
38. Hegedűs, M.; Tóth-Bodrogi, E.; Jónás, J.; Somlai, J.; Kovács, T. Mobility of ^{232}Th and ^{210}Po in red mud. *J. Environ. Radioact.* **2018**, *184–185*, 71–76. [CrossRef] [PubMed]
39. Santos, A.J.G.; Silva, P.S.C.; Mazzilli, B.P.; Fávaro, D.I.T. Radiological characterisation of disposed phosphogypsum in Brazil: Evaluation of the occupational exposure and environmental impact. *Radiat. Prot. Dosim.* **2006**, *121*, 179–185. [CrossRef] [PubMed]
40. Lauer, N.E.; Hower, J.C.; Hsu-Kim, H.; Taggart, R.K.; Vengosh, A. Naturally occurring radioactive materials in coals and coal combustion residuals in the united states. *Environ. Sci. Technol.* **2015**, *49*, 11227–11233. [CrossRef] [PubMed]
41. Ebaid, Y.Y. Use of Gamma-Ray spectrometry for uranium isotopic analysis in environmental samples. *Rom. J. Phys.* **2010**, *55*, 69–74.
42. Ebaid, Y.Y.; El-Mongy, S.A.; Allam, K.A. ^{235}U-γemission contribution to the 186 keV energy transition of ^{226}Ra in environmental samples activity calculations. *Int. Congr. Ser.* **2005**, *1276*, 409–411. [CrossRef]
43. AQAIQ; SAC. *Limits of Radionuclides in Building Materials GB 6566-2010 (In Chinese)*; AQSIQ, SAC: Beijing, China, 2010; pp. 1–3.
44. Lu, X.; Chao, S.; Yang, F. Determination of natural radioactivity and associated radiation hazard in building materials used in Weinan, China. *Radiat. Phys. Chem.* **2014**, *99*, 62–67. [CrossRef]
45. Hamilton, E.I. The relative radioactivity of building materials. *Am. Ind. Hyg. Assoc. J.* **1971**, *32*, 398. [CrossRef]
46. Matthew, P.J. Natural radioactivity of Australian building materials, industrial wastes and by-products. *Health Phys.* **1985**, *48*, 87. [CrossRef]
47. Stojanovska, Z.; Nedelkovski, D.; Ristova, M. Natural radioactivity and human exposure by raw materials and end product from cement industry used as building materials. *Radiat. Meas.* **2010**, *45*, 969–972. [CrossRef]
48. NEA-OECD. *Exposure to Radiation from the Natural Radioactivity in Building Materials: Report by a Group of Exports of the OECD Nuclear Energy Agency*; NEA-OECD: Paris, France, 1979; pp. 13–19.
49. Yu, K.N.; Guan, Z.J.; Stokes, M.J.; Young, E.C.M. The assessment of the natural radiation dose committed to the Hong Kong people. *J. Environ. Radioact.* **1992**, *17*, 31–48. [CrossRef]
50. Hayumbu, P.; Zaman, M.B.; Lubaba, N.C.H.; Munsanje, S.S.; Muleya, D. Natural radioactivity in Zambian building materials collected from Lusaka. *J. Radioanal. Nuclear Chem.* **1995**, *199*, 229–238. [CrossRef]
51. European Commission. Radiological protection principles concerning the natural radioactivity of building materials. Brussels. *Radiat. Prot.* **1999**, *112*, 5–16.
52. He, Z.Y.; Luo, G.Z.; Huang, J.J. National survey on natural radioactivity level of environment (1983–1990) (in Chinese). *Radiat. Prot.* **1992**, *12*, 81–95.
53. Quindos, L.S.; Fernández, P.L.; Ródenas, C.; Gómez-Arozamena, J.; Arteche, J. Conversion factors for external gamma dose derived from natural radionuclides in soils. *J. Environ. Radioact.* **2004**, *71*, 139. [CrossRef]
54. Uosif, M.A.; Eltaher, A. Radiological assessment of Abu-Tartur phosphate, Western Desert Egypt. *Radiat. Prot. Dosim.* **2008**, *130*, 228–235. [CrossRef]
55. Billa, J.; Han, F.; Didla, S.; Ankrah, M.; Yu, H.; Dimpah, J.; Brempong, O.; Adzanu, S. Evaluation of radioactivity levels in fertilizers commonly used in the Southern USA. *J. Radioanal. Nuclear Chem.* **2015**, *306*, 183–191. [CrossRef]

56. UNSCEAR. *Sources and Effects of Ionizing Radiation: United Nations Scientific Committee on the Effects of Atomic Radiation: UNSCEAR 2000 Report to the General Assembly, with Scientific Annexes*; UNSCEAR: New York, NY, USA, 2000; ISBN 9211422388.
57. Uosif, M.A.M.; Mostafa, A.M.A.; Elsaman, R.; Moustafa, E.S. Natural radioactivity levels and radiological hazards indices of chemical fertilizers commonly used in Upper Egypt. *J. Radiat. Res. Appl. Sci.* **2014**, *7*, 430–437. [CrossRef]
58. Ivanovich, M.; Harmon, R.S. *Uranium-Series Disequilibrium: Applications to Earth, Marine, and Environmental Sciences*, 2nd ed.; Clarendon Press and Oxford University Press: Oxford, UK, 1992; p. 571. ISBN 019854278X.
59. El-Bahi, S.M.; Sroor, A.; Mohamed, G.Y.; El-Gendy, N.S. Radiological impact of natural radioactivity in Egyptian phosphate rocks, phosphogypsum and phosphate fertilizers. *Appl. Radiat. Isot.* **2017**, *123*, 121–127. [CrossRef]
60. Megumi, K.; Oka, T.; Yaskawa, K.; Sakanoue, M. Contents of natural radioactive nuclides in soil in relation to their surface area. *JGR Solid Earth* **1982**, *87*, 10857–10860. [CrossRef]
61. Karangelos, D.J.; Anagnostakis, M.J.; Hinis, E.P.; Simopoulos, S.E.; Zunic, Z.S. Determination of depleted uranium in environmental samples by gamma-spectroscopic techniques. *J. Environ. Radioact.* **2004**, *76*, 295–310. [CrossRef]
62. UNSCEAR. *Sources and Effects of Ionizing Radiation, V. I: UNSCEAR 2008 Report to the General Assembly with Scientific Annexes*; United Nations Publication: New York, NY, USA, 2010; ISBN 978-92-1-142274-0.
63. Lederer, J.; Trinkel, V.; Fellner, J. Wide-scale utilization of MSWI fly ashes in cement production and its impact on average heavy metal contents in cements: The case of Austria. *Waste Manag.* **2017**, *60*, 247–258. [CrossRef] [PubMed]
64. Del Valle-Zermeño, R.; Formosa, J.; Chimenos, J.M.; Martínez, M.; Fernández, A.I. Aggregate material formulated with MSWI bottom ash and APC fly ash for use as secondary building material. *Waste Manag.* **2013**, *33*, 621–627. [CrossRef]
65. García Arenas, C.; Marrero, M.; Leiva, C.; Solís-Guzmán, J.; Vilches Arenas, L.F. High fire resistance in blocks containing coal combustion fly ashes and bottom ash. *Waste Manag.* **2011**, *31*, 1783–1789. [CrossRef] [PubMed]
66. Pan, J.R.; Huang, C.; Kuo, J.; Lin, S. Recycling MSWI bottom and fly ash as raw materials for Portland cement. *Waste Manag.* **2008**, *28*, 1113–1118. [CrossRef]
67. Ribeiro, D.V.; Labrincha, J.A.; Morelli, M.R. Effect of the addition of red mud on the corrosion parameters of reinforced concrete. *Cem. Concr. Res.* **2012**, *42*, 124–133. [CrossRef]
68. Senff, L.; Hotza, D.; Labrincha, J.A. Effect of red mud addition on the rheological behaviour and on hardened state characteristics of cement mortars. *Constr. Build. Mater.* **2011**, *25*, 163–170. [CrossRef]

© 2019 by the authors. Licensee MDPI, Basel, Switzerland. This article is an open access article distributed under the terms and conditions of the Creative Commons Attribution (CC BY) license (http://creativecommons.org/licenses/by/4.0/).

MDPI\
St. Alban-Anlage 66\
4052 Basel\
Switzerland\
Tel. +41 61 683 77 34\
Fax +41 61 302 89 18\
www.mdpi.com

Minerals Editorial Office\
E-mail: minerals@mdpi.com\
www.mdpi.com/journal/minerals

www.ingramcontent.com/pod-product-compliance
Lightning Source LLC
LaVergne TN
LVHW071952080526
838202LV00064B/6731